Newnes Digital Logic IC Pocket Book

To my Lady Esther, with love and with gratitude for the endless patience and unfailing kindness that she has shown me throughout the two years that it took me to evaluate and test all of the devices and circuits described in this book and to create the texts that accompany them.

NEWNES

DIGITAL LOGIC IC

POCKET BOOK

Volume 3 in the
Newnes Electronics Circuits
Pocket Book Series

R. M. Marston

Newnes
An imprint of Butterworth-Heinemann

Newnes
An imprint of Butterworth-Heinemann
Linacre House, Jordan Hill, Oxford OX2 8DP

⊄ A member of the Reed Elsevier plc group

OXFORD   BOSTON   JOHANNESBURG
MELBOURNE   NEW DELHI   SINGAPORE

First published 1996
Transferred to digital printing 2004
© R. M. Marston 1996

British Library Cataloguing in Publication Data
A catalogue record for this book is available from the
British Library

ISBN 0 7506 3018 3

Library of Congress Cataloguing in Publication Data
A catalogue record for this book is available from the
Library of Congress

Typeset by 🅐 Tek-Art, Croydon, Surrey

Contents

Preface

Modern digital electronics is dominated by two basic logic technologies, those of TTL (Transistor–Transistor Logic) and CMOS (Complementary MOS-FET logic), and by two major ranges of general-purpose digital ICs, the '74'-series and the '4000'-series. This uniquely informative and practical 'Pocket Book' book takes an in-depth look at both of these technologies and at the most important types of digital IC available in the '74'- and '4000'-series ranges of devices. The book presents the reader with 620 outstandingly useful and carefully selected circuits, diagrams, graphs and tables, backed up by over 75,000 words of highly informative 'how it works' and 'how to use it' text and captions, and gives detailed descriptions and application information on more than 185 modern digital ICs.

The manual is split into 15 chapters. The first three explain digital IC basics, describe TTL and CMOS principles, introduce the various modern sub-families within the '74'-series range of ICs, and explain TTL basic-usage rules, etc. Chapter 4 takes a detailed look at modern CMOS principles and usage rules, and Chapter 5 gives a practical introduction to CMOS digital IC use via the 4007UB IC. Chapter 6 deals with modern logic circuitry; it starts off by looking at the symbology and mathematics of digital logic, then presents a mass of practical logic circuitry and data. The next seven chapters progress through bilateral switch ICs, waveform generator circuitry, clocked flip-flop and counter circuits, special counter/dividers, data latches, registers, comparators, and code converters. Chapter 14 deals with specialized types of IC such as multiplexers, demultiplexers, addressable latches, decoders, full-adders, bus transceivers, priority encoders, and rate multipliers, etc. The final chapter presents a miscellaneous collection of digital IC circuits.

The book, though aimed specifically at all practical design engineers, technicians, and experimenters, will doubtless also be of great interest to all amateurs and students of electronics. It deals with its subject in an easy-to-read, down-to-earth, mainly non-mathematical but very comprehensive and professional manner. Each chapter begins by explaining the basic principles of its subject and then goes on to present the reader with a great mass of practical circuits and useful data, all of which have been fully evaluated and/or verified by the author.

Throughout the volume, great emphasis is placed on practical 'user' information and circuitry, and this book, like all other volumes in the *Electronic Circuits Pocket Book* series, abounds with useful facts and data. Most of the ICs described in the book are modestly priced and readily available types.

The reader may note that I generated all of this manual's 620 diagrams, graphs and tables, etc., via a standard 33MHz 486DX PC and LaserJet IIIp printer, using the excellent low-cost 'Top Draw 2.0' and 'CorelDRAW 3' Windows artwork/CAD packages and my own sets of circuit symbols.

R. M. Marston
1996

1 Digital Logic IC Basics

Modern digital 'logic' ICs are widely available in three basic types, as either TTL devices (typified by the 74LS00 logic family, etc.), as 'slow' CMOS devices (typified by the '4000' logic family), or as 'fast' CMOS devices (typified by the 74HC00 and 74AC00 logic families). Each of these families has its own particular advantages and disadvantages, and its own special set of usage rules. This book sets out to explain the basic principles and usage rules, etc., of each of these three basic digital logic families, and to act as a practical usage guide to the vast range of ICs that are available in each of these families. This opening chapter concentrates on digital logic IC basics.

Digital Logic IC Basics

An IC can be simply described as a complete electronic circuit or 'electronic building block' that is integrated within one or more semiconductor slices or 'chips' and encapsulated in a small multi-pin package, and which can be made fully functional by merely wiring it to a suitable power supply and connecting various pins to appropriate external input, output, and auxiliary networks.

ICs come in both 'linear' and 'digital' forms. Linear ICs are widely used as pre-amplifiers, power amplifiers, oscillators, and signal processors, etc., and give a basic output that is directly proportional to the magnitude (analogue value) of the input signal, which itself may have any value between zero and some prescribed maximum limit. One of the simplest types of linear IC element is the unity-gain buffer; if a large sine-wave signal is connected to the input of this circuit, it produces a low-impedance output of almost identical form and amplitude, as shown in *Figure 1.1(a)*. Digital ICs, on the other hand, are effectively blind to the precise amplitudes of their input signals, and simply recognize them as being in either a 'low' or a 'high' state (usually known as 'logic-0' and 'logic-1' states

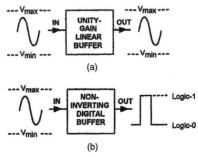

Figure 1.1   When a large input sine wave is fed to the input of a linear buffer *(a)*, it produces a good sine-wave output, but when fed to the input of a digital buffer *(b)* it produces a purely digital output

respectively); their outputs similarly have only two basic states, either 'low' or 'high' (logic-0 or logic-1). One simple type of digital IC element is the non-inverting buffer; if a large sine-wave signal is connected to the input of this circuit, it produces an output that (ideally) is of purely digital form, as shown in *Figure 1.1(b)*.

Digital ICs are available in a variety of rather loosely defined categories such as 'memory' ICs, 'electronic delay-line' ICs, and 'microprocessor support' ICs, etc., but the most widely used category is that known as the 'logic' type, in which the ICs are designed around fairly simple logic circuits such as digital buffers, inverters, gates, or flip-flop elements. Digital logic *circuits* come in a variety of basic types, and can be built using a variety of types of discrete or integrated technologies; *Figures 1.2* to *1.7* show a selection of very simple logic circuits that are designed around discrete components.

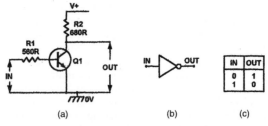

Figure 1.2   Circuit *(a)*, symbol *(b)*, and Truth Table *(c)* of a simple inverting digital buffer

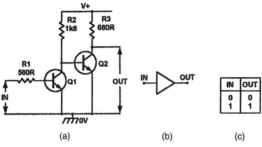

Figure 1.3 Circuit *(a)*, symbol *(b)*, and Truth Table *(c)* of a non-inverting digital buffer

Figure 1.2(a) shows a simple inverting digital buffer (also known as a NOT logic gate), consisting of an unbiased transistor wired in the common-emitter mode, and *(b)* shows the international symbol that is used to represent it (the arrow-head indicates the direction of signal flow, and the small circle on the symbol's output indicates the 'inverting' action). The circuit action is such that Q1 is cut off (with its output high) when its input is in the zero state, and is driven fully on (with its output pulled low) when its input is high; this information is presented in concise form by the Truth Table of *(c)*, which shows that the output is at logic-1 when the input is at logic-0, and vice versa.

Figure 1.3(a) shows a simple non-inverting digital buffer, consisting of a direct coupled pair of common-emitter (inverter) transistor stages, and *(b)* shows the arrow-like international symbol that is used to represent it; *(c)* shows the Truth Table that describes its action, e.g. the output is at logic-0 when the input is at logic-0, and is at logic-1 when the input is at logic-1.

In digital electronics, a 'gate' is a logic circuit that opens or gives an 'output' (usually defined as a 'high' or logic-1 state) only under a certain set of input conditions. *Figure 1.4(a)* shows a simple 2-input OR gate, made from two diodes and a resistor, and *(b)* shows the international symbol that is used to represent it; *(c)* shows its Truth Table (in which the inputs are referred to as A or B), which shows that the output goes to logic-1 if A *or* B goes to logic-1.

Figure 1.5 shows the circuit, symbol, and Truth Table of a 2-input NOR (Negated-output OR) gate, in which the

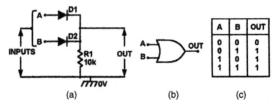

A	B	OUT
0	0	0
0	1	1
1	0	1
1	1	1

(a)   (b)   (c)

Figure 1.4   Circuit *(a)*, symbol *(b)*, and Truth Table of a simple 2-input OR gate

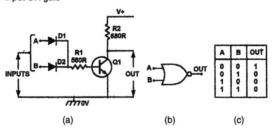

A	B	OUT
0	0	1
0	1	0
1	0	0
1	1	0

(a)   (b)   (c)

Figure 1.5   Circuit *(a)*, symbol *(b)*, and Truth Table of a 2-input NOR gate

output is inverted (as indicated by the output circle) and goes to logic-0 if either input goes high.

Figure 1.6(a) shows a simple 2-input AND gate, made from two diodes and a resistor, and *(b)* shows its standard international symbol; *(c)* shows the AND gate's Truth Table, which indicates that the output goes to logic-1 only if inputs A *and* B are at logic-1.

Finally, *Figure 1.7* shows the circuit, symbol, and Truth Table of a 2-input NAND (Negated-output AND) gate, in which the output is inverted (as indicated by the

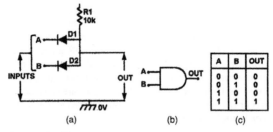

A	B	OUT
0	0	0
0	1	0
1	0	0
1	1	1

(a)   (b)   (c)

Figure 1.6   Circuit *(a)*, symbol *(b)*, and Truth Table of a simple 2-input AND gate

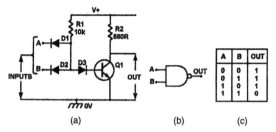

Figure 1.7 Circuit *(a)*, symbol *(b)*, and Truth Table of a 2-input NAND gate

output circle) and goes to logic-0 only if both inputs are at logic-1.

Note that although the four basic types of logic gate circuit described above are each shown with only two input terminals, they can in fact be designed or used to accept any desired number of inputs, and can be used to perform a variety of simple 'logic' operations. In practice, many types of digital buffer and gate are readily available in IC form, as also are many other digital logic circuits, including flip-flops, latches, shift registers, counters, data selectors, encoders, and decoders, etc.

Practical 'digital' ICs may range from relatively simple 'logic' devices housing the equivalent of just a few basic gates or buffers, to incredibly complex devices housing the equivalent of tens of thousands of interconnected gates, etc. By convention, the following general terms are used to describe the relative density or complexity of integration:

- *SSI (Small Scale Integration)*; complexity level between 1 and 10 gates.

- *MSI (Medium Scale Integration)*; complexity level between 10 and 100 gates.

- *LSI (Large Scale Integration)*; complexity level between 100 and 1000 gates.

- *VLSI (Very Large Scale Integration)*; complexity level between 1000 and 10,000 gates.

- *SLSI (Super Large Scale Integration)*; complexity level between 10,000 and 100,000 gates.

Note that most 'logic' ICs of the types described throughout this book have complexity levels ranging from 4 to 400 gates, and are thus SSI, MSI, or LSI devices. In general terms, most microprocessor ICs and moderately large memory ICs are VLSI devices, and large dynamic RAM (Random Access Memory) ICs are SLSI devices.

Digital Waveform Basics

Digital logic ICs are invariably used to process digital waveforms. It is thus pertinent at this point to remind the reader of some basic facts and terms concerning digital waveforms, which are available in either square or pulse form. *Figure 1.8* illustrates the basic parameters of a squarewave; in each cycle the wave first switches from zero to some peak voltage value (V_{pk}) for a fixed period, and then switches low again for a second fixed period, and so on. The time taken for the waveform to rise from 10% to 90% of V_{pk} is known as its *rise time*, and that taken for it to drop from 90% to 10% of V_{pk} is known as its *fall time*. In each squarewave cycle, the 'high' part is known as its *mark* and the 'low' part as its *space*. In a symmetrical squarewave such as that shown, the mark and space periods are equal and the waveform is said to have a 1:1 Mark–Space (or M–S) ratio, or a 50% duty cycle (since the mark duration forms 50% of the total cycle period). Squarewaves are not necessarily symmetrical, but are always free-running or repetitive, i.e. they cycle repeatedly, with sharply defined mark and space periods.

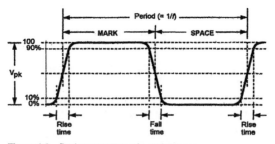

Figure 1.8 Basic parameters of a squarewave

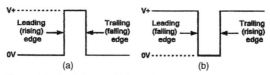

Figure 1.9 Basic forms of *(a)* 'positive-going' and *(b)* 'negative-going' pulses

A pulse waveform can be roughly described as being a bit like a squarewave (complete with rise and fall times, etc.) but with only its mark or its space period sharply defined, the duration of the remaining period being unimportant. *Figure 1.9(a)* shows a basic 'positive-going' pulse waveform, which has a 'rising' or positive-going leading edge, and *(b)* shows a 'negative-going' pulse waveform, which has a 'falling' or negative-going leading edge. Note that many modern MSI digital ICs such as counter/dividers and shift registers, etc., can be selected or programmed to trigger on either the rising or the falling edge of an input pulse, as desired by the user.

If a near-perfect pulse waveform is fed to the input of a real-life amplifier or logic gate, etc., the resulting output waveform will be distorted both in form and time, as shown in *Figure 1.10*. Thus, not only will the output waveform's rise and fall times be increased, but the

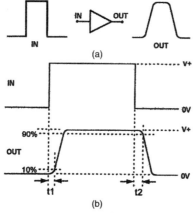

Figure 1.10 A perfect pulse, fed to the input of a practical amplifier or gate, produces an output pulse that is distorted both in form and time; the output pulse's time delay is called its 'propagation delay', and (in *(b)*) = $(t_1+t_2)/2$

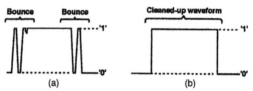

Figure 1.11   Mechanically derived pulse waveforms often suffer from 'contact bounce' *(a)*, and must be cleaned up *(b)* before use

arrival and termination of the output pulse will be time-delayed relative to that of the input pulse; the mean value of the delays is called the device's *propagation delay*. Also, the peaks of the waveform's rising and falling edges may suffer from various forms of 'ringing' or 'overshoot' or 'undershoot'; the magnitudes of all of these distortions varies with the quality or structure of the amplifier or gate, etc.

In practice, pulse input waveforms may sometimes be so imperfect that they may need to be 'conditioned' before they are suitable for use by modern fast-acting digital ICs. Specifically, they may have such long rise or fall times that they may have to be sharpened up via a Schmitt trigger before they are suitable for use. Again, many mechanically derived 'pulse' waveforms such as those generated via switches or contact-breakers, etc., may suffer from severe multiple 'contact bounce' problems such as those shown in *Figure 1.11(a)*, in which case they will have to be converted to the 'clean' form shown in *(b)* before they can be usefully used.

Basic Logic-IC Families

Practical digital logic circuits and ICs can be built by using various technologies. The first successful family of digital logic ICs appeared in the mid 1960s; these used a 3.6V supply and employed a simple technology that became known as Resistor–Transistor Logic, or RTL. *Figure 1.12* shows the basic circuit of a 3-input RTL NOR gate. RTL was rather slow in operation, having a typical propagation delay (the time taken for a single pulse edge or transition to travel from input to

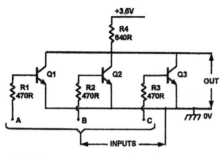

Figure 1.12   IC version of a 3-input RTL NOR gate

output) of 40ns in a low-power gate, or 12ns in a medium-power gate. RTL is now obsolete.

Another early type of IC logic technology, developed in the late 1960s, was based on simple developments of the discrete types of logic circuit shown in *Figures 1.2* to *1.7*, and was known as Diode–Transistor Logic, or DTL. *Figure 1.13* shows the basic circuit of a 3-input DTL NAND gate. DTL used a dual 5 volt power supply, gave a typical propagation delay of 30ns, and gave an output of less than 0.4V in the '0' state and greater than 3.5V in the '1' state. DTL is now obsolete.

Between the late 1960s and mid 1970s, several other promising IC logic technologies appeared, most of them soon disappearing back into oblivion. Among those that came and either went or receded in importance were HTL (High Threshold Logic), ECL (Emitter Coupled Logic), and PML (P-type MOSFET Logic). The most durable of these technologies was ECL, which is still in production and gives very fast operation, but at the cost of very high current/power consumption.

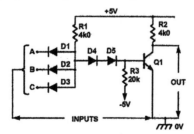

Figure 1.13   IC version of a 3-input DTL NAND gate

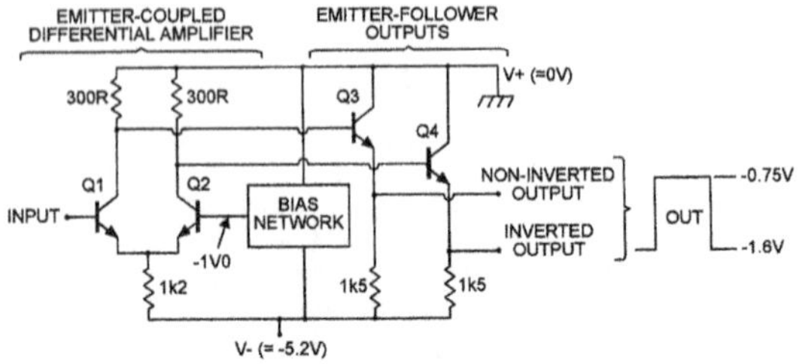

Figure 1.14 Basic ECL (Emitter-Coupled Logic) amplifier circuit

Figure 1.14 shows the basic circuit of the ECL digital amplifier, which is simply a non-saturating emitter-coupled differential amplifier (Q1 and Q2) with emitter–follower output stages (Q3 and Q4), and – because its transistors do not saturate when switched – gives typical propagation delays of only 4ns. Note that the circuit's V+ line is normally grounded and the V- line is powered at –5.2V, and under this condition the circuit provides nominal digital output swings of only 0.85V, i.e. from a low state of –1.60V to a high state of –0.75V. The circuit's digital input is applied to the base of Q1, and a non-inverted output is available on Q3 emitter, and an inverted output is available on Q4 emitter. Modern ECL ICs are used only when ultra-high-speed operation is required.

The basic aim of digital IC designers during the late 1960s to early 1970s period was that of devising a technology that was simple to use and which gave a good compromise between high operating speed and low power consumption. The problem here was that conventional transistor-type circuitry, using an output stage of the *Figure 1.2* type (as in RTL and DTL systems, etc.) was simply not capable of meeting the last two of these design needs. The essence of this problem – and its ultimate solution – can be understood with the aid of *Figure 1.15*.

Figure 1.15(a) shows a simplified version of the basic *Figure 1.2* circuit, with Q1 replaced by a simple mechanical switch. Remember here that all practical output loads inevitably contain capacitance (typically up to about 30pF in most digital circuits), so it can be seen that this basic circuit will charge (source current into) a

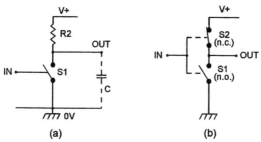

Figure 1.15 *(a)* Simple digital switch. *(b)* Basic 'totem-pole' digital switch

capacitive load fairly slowly via R2 when S1 is open, but will discharge it (sink current from it) rapidly via S1 when S1 is closed; thus, circuits of this type produce digital outputs that tend to have long rise times and short fall times; the only way to reduce the rise time is to reduce the R2 value, and that increases S1's (Q1's) current consumption by a proportionate amount.

Note that one good way of describing the deficiency of the *Figure 1.15(a)* logic circuit is to say that its output gives an active pull-down action (via S1), but a passive pull-up action (via R2). Obviously, a far better digital output stage could be made by replacing R2 and S1 with a ganged pair of changeover switches, connected as shown in *Figure 1.15(b)*, so that S2 gives active pull-up action and S1 gives active pull-down action, but so arranged that only one switch can be closed at a time (thus ensuring that the circuit consumes zero quiescent current); such a circuit – with one electronic switch placed above the other – is generally known as a 'totem-pole' output stage.

Throughout the late 1960s, digital engineers strove to design a cheap and reliable electronic version of the totem-pole output stage, and then – in the early 1970s – two such technologies hit the commercial market like a bombshell, and went on to form the basis of today's two dominant digital IC families. The first of these based on bipolar transistor technology, is known as TTL (Transistor–Transistor Logic), and formed the basis of the so-called '74' family of digital ICs that first arrived in 1972. The second based on FET technology, is known as CMOS (*C*omplementary *MOS*-FET logic), and forms the basis of the rival '4000-series' (and the similar '4500-series') digital IC family that first arrived in about 1975. The TTL and CMOS technologies have vastly different characteristics, but both offer specific technical advantages that make them invaluable in particular applications.

The most significant differences between the technologies of CMOS and TTL ICs can be seen in their basic inverter/buffer networks, which are used (sometimes in slightly modified form) in virtually every IC within the family range of each type of device. *Figures 1.16* and *1.17* show the two different basic designs.

The CMOS inverter of *Figure 1.16* consists of a complementary pair of MOSFETs, wired in series, with p-channel MOSFET Q1 at the top and n-channel MOSFET Q2 below, and with both high-impedance gates joined together. The pair can be powered from any supply in the 3V to 15V range. When the circuit's input is at logic-0, the basic action is such that Q1 is

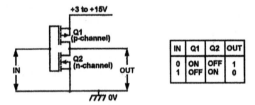

IN	Q1	Q2	OUT
0	ON	OFF	1
1	OFF	ON	0

Figure 1.16   Circuit and Truth Table of a basic CMOS inverter

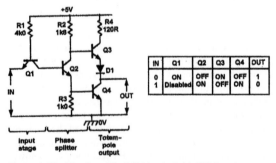

IN	Q1	Q2	Q3	Q4	OUT
0	ON	OFF	ON	OFF	1
1	Disabled	ON	OFF	ON	0

Figure 1.17   Circuit and Truth Table of a basic TTL inverter

driven on and Q2 is cut off, and the output is actively pulled high (to logic-1); note that the output can source (drive) fairly high currents into an external load (via Q1) under this condition, but that the actual inverter stage consumes near-zero current, since Q2 is cut off. When the circuit's input is at logic-1, the reverse of this action occurs, and Q1 is cut off and Q2 is driven on, and the output is actively pulled low (to logic-0); note that the output can sink (absorb) fairly high currents from an external load (via Q2) under this condition, but that the actual inverter stage consumes near-zero current, since Q1 is cut off.

Thus, the basic CMOS inverter can be used with any supply in the 3V to 15V range, has a very high input impedance, consumes near-zero quiescent current, has an output that switches almost fully between the two supply rails, and can source or sink fairly high output load currents. Typically, a single basic CMOS stage has a propagation delay of about 12 to 60ns, depending on supply voltage.

The TTL inverter of *Figure 1.17* is split into three sections, consisting of an emitter-driven input (Q1), a phase-splitter (Q2), and a 'totem-pole' output stage (Q3–D1–Q4), and must be powered from a 5V supply. When the circuit's input is pulled down to logic-0, the basic action is such that Q1 is saturated, thus depriving Q2 of base current and causing Q2 and Q4 to cut off, and at the same time causing emitter–follower Q3 to turn on via R2 and give an active pull-up action in which the output has (because of various volt-drops) a typical loaded value of about 3.5V and can source fairly high currents into an external load. Conversely, when the circuit's input is at logic-1, Q1 is disabled, allowing Q2 to be driven on via R1 and the forward-biased base-collector junction of Q1, thus driving Q4 to saturation and simultaneously cutting off Q3; under this condition Q4 gives an active pull-down action and can sink fairly high currents, and the output takes up a typical loaded value of 400mV. Note that (ignoring external load currents) the circuit consumes a quiescent current of about 1mA in the logic-1 output state, and 3mA in the logic-0 output state.

Thus, the basic TTL inverter can only be used with a 5V supply, has a very low input impedance, consumes up to 3mA of quiescent current, has an output that does not switch fully between the two supply rails, and can source or sink fairly high load currents. Typically, a single basic TTL stage has a propagation delay of about 12ns.

Basic TTL Circuit Variations

There are five very important variations of the basic *Figure 1.17* TTL 'inverter' circuit. The simplest of these is the so-called 'open collector' TTL circuit, which is shown in basic form in *Figure 1.18*. Here, output transistor Q3 is cut off when the input is at logic-0, and is driven on when the input is at logic-1. Thus, by wiring an external load resistor between the 'OUT' and '+5V' pins, the circuit can be used as a 'passive pull-up' voltage inverter that has an output that (when lightly loaded) switches almost fully between zero and the positive supply rail value. Alternatively, it can be used to drive an external load (such as an LED or relay, etc.) that is connected between 'OUT' and a positive supply rail, in which case the load activates when a logic-1 input is applied.

The second variation is the non-inverting amplifier or buffer. This is made by simply wiring an additional direct-coupled inverter stage between the phase-splitter and output stages of the standard inverter.

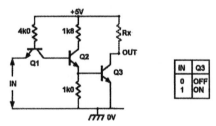

Figure 1.18   TTL inverter with open-collector output

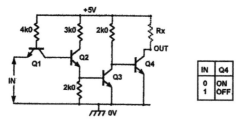

Figure 1.19 TTL non-inverting buffer with open-collector output

Figure 1.19 shows an 'open collector' version of such a circuit, which can be used with an external resistor or load; in this example, Q4 turns on when a logic-0 input is applied.

Figure 1.20 shows a major TTL design variation. Here, the basic inverter circuit is used with a triple-emitter input transistor, to make a 3-input NAND gate in which the output goes low (to logic-0) only when all three inputs are high (in the logic-1 state). In practice, multiple-emitter transistors are widely used within TTL ICs; some TTL gates use an input transistor with as many as a dozen emitters, to make a 12-input gate.

A further variation concerns the use of a 'Tri-State' or '3-state' type of output that incorporates additional networks plus an external ENABLE control terminal, which in one state allows the totem-pole output stage to operate in its normal 'logic-0 or logic-1' mode, but in the other state disables (turns off) both totem-pole transistors and thus gives an open-circuit (high impedance) output. This facility is useful in allowing several outputs or inputs to be shorted to a common bus

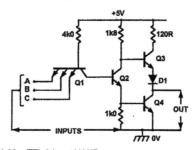

Figure 1.20 TTL 3-input NAND gate

or line, as shown in *Figure 1.21*, and to communicate along that line by ENABLING only one output and one input device at a time.

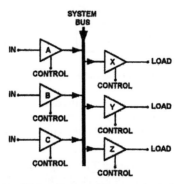

Figure 1.21   'Tri-State' logic enables several outputs or inputs to be connected to a common bus; only one output/input must be made active at any given moment

The final circuit variation is an 'application' one, and concerns the use of an external 2k2 'pull-up' resistor on a totem-pole output stage, as shown in *Figure 1.22*. This resistor pulls the output (when lightly loaded) up to virtually the full +5V supply value when the output is in the logic-1 state, rather than to only +3.5V; this is sometimes useful when interfacing the output of a TTL IC to the input of a CMOS IC, etc.

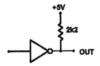

Figure 1.22   An external 2k2 pull-up resistor connected to the output of a totem-pole stage pulls the output to almost +5V in the logic-1 state

The '74'-series of Digital ICs

TTL IC technology first hit the electronics engineering scene in a big way in about 1972, when it arrived in the form of an entire range of digital logic ICs that were exceptionally easy to use. The range was an instant international success, and quickly became the world's leading IC logic system. Its ICs were produced in both commercial and military grades, and carried the prefixes of '74' and '54' respectively; the commercial product range soon became known simply as the '74'-series of ICs.

Over the years, the '74'-series of ICs has progressively expanded its range of devices and advanced its production technology, so that today the '74'-series is as popular and versatile as ever. When first introduced in the early 1970s, the series was based entirely on a simple type of TTL technology, but in later years new sub-families of TTL were introduced in the series, and then various types of CMOS technology were added to it, so that today's '74-series' incorporates a variety of TTL and CMOS sub-families. Chapter 2 takes a close look at the '74' sub-families of ICs, and explains some basic TTL terminology.

2   The '74'-series of Digital ICs

Chapter 1 started off by looking at digital logic IC basics, went on to describe basic logic IC families – including TTL and CMOS – and concluded by introducing the '74'-series of ICs. The present chapter takes a deeper look at the '74'-series of ICs, and at various sub-members of the TTL and CMOS families.

The '74'-series of Digital ICs

Modern digital electronics is dominated by two major logic IC families, these being the low-speed '4000'-series of CMOS ICs, and the '74'-series of fast TTL and CMOS ICs. The '74' family was originally based entirely on TTL technology, which first hit the electronics scene in a big way in about 1972, when the '74'-series suddenly arrived in the form of an entire range of versatile and cleverly conceived TTL digital logic ICs that were each designed to operate from a single-ended 5-volt supply and to directly and easily interconnect with each other without hassle (each output could directly drive several inputs), thus making it relatively easy for any moderately competent engineer to design and develop fairly complex digital logic systems. The series was an instant and brilliant international success, and almost immediately became the world's leading IC logic system. Its ICs were produced in both commercial and military grades, and carried prefixes of '74' and '54' respectively; the commercial product range rapidly became known simply as the '74'-series of ICs.

A major feature of the '74'-series is that all devices within the range function as 'black boxes' that operate at similar input and output threshold levels; the user does not need to understand their internal circuitry in order to use them, but simply needs to know their basic usage rules. Also, the input sensitivity or 'fan-in' of each device conforms to a fixed standard, and its output drive capability or 'fan-out' has a guaranteed minimum value that indicates the number of external '74'-series inputs that it can safely directly drive, making it very easy to interconnect various devices; thus, the output of a '74' gate with a fan-out of 10 can directly drive as many as ten parallel-connected standard inputs on other '74'-series ICs.

The type of TTL technology used in the initial (1972) '74'-series ICs resulted in a range of devices that were moderately fast but consumed fairly heavy currents. Within a year or so, sub-families of the original TTL were introduced, offering a trade-off between speed and power, i.e. twice the speed but at twice the current consumption (in the 'H' or 'high speed' sub-family), or one-tenth of the current consumption but only one-third of the normal speed (in the 'L' or 'low power' sub-family), etc. This trend of seeking a good or ever-better trade-off between speed and power consumption has continued until the present day, and so far a total of eight commercially successful sub-families of TTL (and five sub-families of CMOS) have appeared in the '74'-series of digital ICs. Many of these sub-families have subsequently become obsolete or obsolescent, but the practical design/maintenance engineer or technician still needs a basic knowledge of all of them, since they are often found in old equipment that needs repairing or upgrading.

Note that each sub-family of the '74'-series of ICs is almost directly compatible with all other sub-families in the series. Thus, if you open up an old piece of equipment and find that an old '74L90' decade counter IC needs replacing but is no longer available, you will probably find that a modern '74LS90' decade counter IC can be used as a direct plug-in replacement, or that any other '74XX90' IC can be used as a replacement either directly or with slight circuit modification (depending on the basic characteristics of the two sub-families). In either case, the first thing that you will need to do is identify the device of interest, from its printed code number. *Figure 2.1* explains the basic scheme that is used in formating the '74'-series code numbers.

All ICs in the '74' family are identified by an alpha-numeric code which, in its simplest form, consists of three

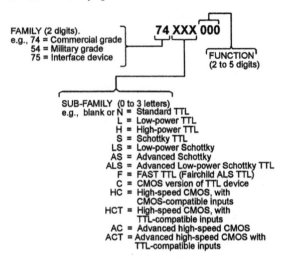

FAMILY (2 digits).
e.g., 74 = Commercial grade
54 = Military grade
75 = Interface device

74 XXX 000

FUNCTION
(2 to 5 digits)

SUB-FAMILY (0 to 3 letters)
e.g., blank or N = Standard TTL
L = Low-power TTL
H = High-power TTL
S = Schottky TTL
LS = Low-power Schottky
AS = Advanced Schottky
ALS = Advanced Low-power Schottky TTL
F = FAST TTL (Fairchild ALS TTL)
C = CMOS version of TTL device
HC = High-speed CMOS, with
CMOS-compatible inputs
HCT = High-speed CMOS, with
TTL-compatible inputs
AC = Advanced high-speed CMOS
ACT = Advanced high-speed CMOS with
TTL-compatible inputs

Figure 2.1   Basic coding system used on the '74'-series ICs

sub-codes strung together as shown in *Figure 2.1*. The first (left-hand) sub-code consists of 2 digits that read either 74, 54, or 75. '74' identifies the IC as a commercial-grade member of the family; these devices are usually encapsulated in a plastic 14-pin, 16-pin, or 24-pin dual-in-line package (DIP), can be used with supplies within the limit +4.75V to +5.25V, and can be operated over the temperature range 0°C to +70°C. '54' identifies the IC as a high-quality military-grade member of the family; these devices are encapsulated in exotic packages, can use supplies within the limit +4.5V to +5.5V, and can operate over the temperature range -55°C to +125°C. '75' identifies the IC as a commercial-grade interface device that is designed to support the '74' range of devices.

The second (central) sub-code consists of up to three letters, and identifies the precise technology or sub-family used in the construction of the device, as shown in the diagram. Note that standard TTL devices carry either no central code at all, or an 'N'; each of the other seven major TTL sub-family devices carry a central identifying code, and the five major CMOS '74' sub-families carry a central code that includes the letter 'C'.

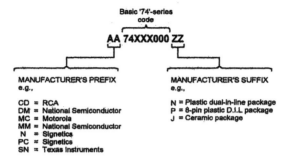

Basic '74'-series
code

AA 74XXX000 ZZ

MANUFACTURER'S PREFIX e.g.,	MANUFACTURER'S SUFFIX e.g.,
CD = RCA	N = Plastic dual-in-line package
DM = National Semiconductor	P = 8-pin plastic D.I.L package
MC = Motorola	J = Ceramic package
MM = National Semiconductor	
N = Signetics	
PC = Signetics	
SN = Texas Instruments	

Figure 2.2 The basic '74'-series code is often elaborated with a manufacturer's prefix and/or suffix

The last (right-hand) sub-code usually consists of 2 to 5 digits (but occasionally includes a letter 'A' or a star), and identifies the precise function of the IC (e.g. quad 2-input NAND gate, decade counter, 4-bit shift register, etc.). The precise relationship between this sub-code and the device function can be ascertained from manufacturers' lists.

Thus, a '74' type of IC may carry a code that, in its simplest form, reads something like 7400, 74N00, or 7414, etc., if it is a standard TTL device, or 74L14, 74LS38, or 74HC03, etc., if it is some other sub-member of the '74' family. Note that in practice '74'-series ICs often carry an elaborated form of the basic code that includes a two-letter prefix that identifies the manufacturer, plus a lettered manufacturer's suffix that indicates the packaging style, etc., as shown in *Figure 2.2*. Hence, a device marked SN74LS90N is a normal 74LS90 IC, manufactured by Texas Instruments and housed in a plastic dual-in-line package.

TTL Sub-families

Eight major sub-families of TTL have been used in the '74'-series throughout its lifetime, as follows:

Standard TTL. Standard TTL is similar to the basic type already described, except that each of its inputs is provided with a protection diode that helps suppress transients and speed up its switching action. *Figure 2.3* shows the actual circuit of a 7400 2-input NAND gate; its power consumption is 10mW, and its propagation delay is 9ns when driving a 15pF/400R load.

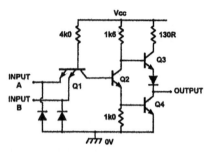

Figure 2.3   Circuit of a Standard TTL 7400 2-input NAND gate

Low-power (L) TTL (now obsolete). Low-power TTL is a modified version of the standard type, with its resistance values greatly increased to give a dramatic reduction in power consumption at the expense of reduced speed. *Figure 2.4* shows the circuit of a 74L00 2-input NAND gate; its power consumption is 1mW, and its propagation delay is typically 33ns.

High-speed (H) TTL (now obsolete). High-speed TTL is a modified version of the standard type, with its resistance values reduced to give an increase in speed at

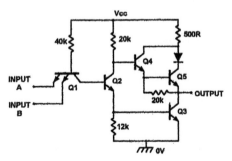

Figure 2.4   Circuit of a low-power (L) TTL 74L00 2-input NAND gate

the expense of increased power consumption. *Figure 2.5* shows the circuit of a 74H00 2-input NAND gate; its power consumption is 22mW, and its propagation delay is typically 6ns.

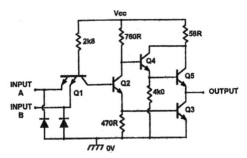

Figure 2.5   Circuit of a high-speed (H) TTL 74H00 2-input NAND gate

Schottky (S) TTL (now obsolete). A transistor switch can be designed to give either a saturated or an unsaturated switching action. Saturated switching – in which the transistor's collector voltage falls far below that of the base under the 'on' condition – is very easy to implement, but produces propagation delays that are about 2.5 times longer than those available from unsaturated circuits. Standard TTL operates its transistors in a heavily saturated switching mode in which the collector falls some 400mV below the base under the 'on' condition, and is thus intrinsically fairly slow. Schottky TTL, on the other hand, operates its transistors in a lightly saturated switching mode in which the collector only falls some 180mV below the base voltage under the 'on' condition, and is almost as fast as an unsaturated circuit (such as an ECL design). Basically, this action is achieved by connecting a Schottky diode (which is fast acting and has a typical forward volt drop of only 180mV) between the transistor's collector and base as shown in *Figure 2.6(a)*, in which R_s represents the input pulse's source impedance. Thus, if the collector goes more than 180mV negative to the base, the Schottky diode becomes forward biased and starts to shunt base current directly into the transistor's collector, thus automatically

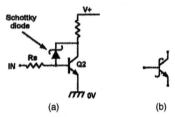

(a)   **(b)**

Figure 2.6   *(a)* a Schottky diode used to limit the saturation depth of an npn transistor; *(b)* symbol of an npn 'Schottky' transistor, with a built-in clamping diode between its collector and base

preventing deeper saturation. In reality, the Schottky diode can easily be incorporated in the transistor's structure, and a 'Schottky-clamped transistor' of this type uses the symbol shown in *(b)*.

In a practical Schottky TTL IC, Schottky-clamped transistors are widely used, and most resistance values are reduced, thus giving a good increase in speed at the expense of power consumption. The totem-pole output stage uses a Darlington transistor pair to give active pull-up, plus a modified active pull-down network that gives an improved waveform-squaring action. *Figure 2.7* shows the circuit of a 74S00 2-input NAND gate; its power consumption is 20mW and its propagation delay is 3ns when driving a 15pF/280R load.

Low-power Schottky (LS) TTL. Low-power Schottky uses a modified form of Schottky technology, using

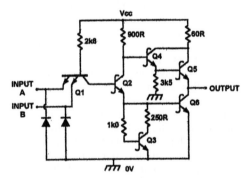

Figure 2.7   Circuit of a Schottky (S) TTL 74S00 2-input NAND gate

improved manufacturing techniques, combined with a 'diode-transistor' (rather than multi-emitter) form of input network that has a high impedance and gives fast switching. *Figure 2.8* shows the circuit of a 74LS00 2-input NAND gate; its power consumption is 2mW and its propagation delay is 8ns when driving a 12p/2k0 load.

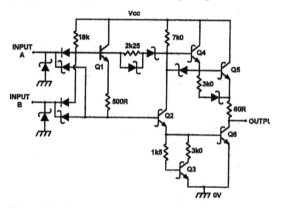

Figure 2.8   Circuit of a low-power Schottky (LS) TTL 74LS00 2-input NAND gate

Advanced low-power Schottky (ALS) TTL. This sub-family is similar to LS but uses an advanced fabrication process which, combined with minor design modifications, yields active devices that are faster and have higher gains than LS types. *Figure 2.9* shows the

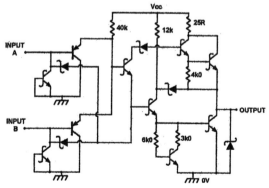

Figure 2.9   Circuit of an advanced low-power Schottky (ALS) 74ALS00 2-input NAND gate

circuit of a 74ALS00 2-input NAND gate; its power consumption is 1mW and its propagation delay is 4ns when driving a 50p/2k0 load.

Advanced Schottky (AS) TTL. This sub-family is similar to ALS, but its design is optimized to give very high speed at the expense of power consumption. *Figure 2.10* shows the circuit of a 74AS00 2-input NAND gate; its power consumption is 22mW and its propagation delay is a mere 2ns when driving a 50p/2k0 load.

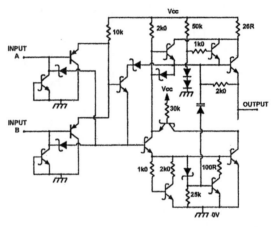

Figure 2.10   Circuit of an advanced Schottky (AS) TTL 74AS00 2-input NAND gate

FAST (F) TTL. FAST (Fairchild Advanced Schottky TTL) is Fairchild's version of 'AS' TTL. It is manufactured under licence by several companies (including Philips and National Semiconductor); its performance is similar (in terms of speed and power consumption) to that of the AS sub-family.

CMOS '74'-series Sub-families

When the '74'-series of ICs was first released in 1972, all devices in the range were based on bipolar TTL technology, which inherently consumes a fairly large amount of power irrespective of its operating speed. In about 1975 the rival CMOS digital ICs technology arrived on the scene, and although not as fast as TTL it offered the outstanding advantage of having a power consumption that was directly proportional to operating speed, being virtually zero under quiescent conditions and rising to the same value as TTL at about 10MHz. In the late 1970s CMOS was introduced as a sub-family within the '74'-series range of devices, carrying the central code 'C'; the graph of *Figure 2.11* compares the frequency/current-consumption curves of a single gate from the standard TTL and the CMOS 'C' versions of the 7400 quad 2-input NAND gate IC.

In its early form, the '74'-series 'C' sub-family was slow and had a very weak output-drive capability (its fan-out drive was equal to two 'L'-type inputs). In subsequent years, however, considerable improvements took place in both the design and production of CMOS-type devices; the salient details of this subject are dealt with in greater detail in Chapter 4, but in the meantime it is sufficient to know that a total of five CMOS sub-families have been introduced in the '74'-series, as follows.

Standard (C) CMOS (now obsolete). This sub-family was virtually normal CMOS in a '74'-series format.

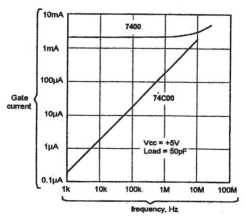

Figure 2.11 Frequency–current graphs of 7400 (TTL) and 74C00 (CMOS) 2-input NAND gates (with a squarewave input)

Typically, a single 74C00 2-input NAND gate consumed about 15mW at 10MHz, and had a propagation delay of 60ns.

High-speed (HC) CMOS. In the early 1980s, advances in CMOS fabrication techniques yielded speed performances similar to LS TTL, but with CMOS levels of power consumption. HC '74'-series devices using this technology have CMOS-compatible inputs; typically, a single 74HC00 2-input NAND gate consumes less than $1\mu A$ of quiescent current, and has a propagation delay of 8ns.

High-speed (HCT) CMOS. These are HC-type devices, but have TTL-compatible inputs. Typically, a 74HCT00 2-input NAND gate consumes less than $1\mu A$ of quiescent current and has a propagation delay of 18ns.

Advanced high-speed (AC) CMOS. In the late 1980s, advances in CMOS design and further advances in CMOS fabrication techniques yielded speed performance similar to those of ALS. AC '74'-series devices using this technology have CMOS-compatible inputs; typically, a 74AC00 2-input NAND gate has a propagation delay of 5ns.

Advanced high-speed (ACT) CMOS. These are AC-type devices, but have TTL-compatible inputs. Typically, a 74ACT00 2-input NAND gate has a propagation delay of 7ns.

Which Logic Family is Best?

Two major general-purpose logic families are currently available, these being the '4000'-series low-speed CMOS family, and the high-speed '74'-series TTL/CMOS family (a third family, using ECL technology, is very specialized and intended for use mainly in very-high-speed applications). The '4000'-series (described in Chapter 4) is of particular value in circuits operating below frequencies of a few MHz in which a minimal figure of quiescent current consumption is desired; other major advantages of the series are that its ICs can operate from any supply in the 3 to 15V range, have excellent noise immunity, and have ultra-high input impedances.

The '74'-series is of special value in circuits operating at frequencies up to several tens of MHz, in which low quiescent current consumption is not too important and in which the ICs can be powered from a well-regulated DC supply (typically of +5V). If you decide to use a '74'-series IC, you are next faced with the problem of deciding 'which sub-family is best for my application?'

Which '74'-series Sub-family is Best?

When designing a new logic circuit, ICs should always be selected on a basis of commercial – rather than purely technical – superiority. It would, for example, be foolish to use a really fast ALS gate in an application in which a slower LS or HC device would be perfectly adequate and was easily available at a fraction of the cost of the ALS device. With this point in mind, note that the five '74'-series IC sub-families most widely available at the time of writing are Standard and LS TTL, and HC, HCT and AC CMOS. Of these, Standard TTL is technically and commercially inferior to LS and is not recommended for use in new designs; AC CMOS cost approximately 2.5 times as much as LS TTL or HC/HCT CMOS and should thus only be used in special applications; and HCT is only meant to be used as a replacement for TTL devices in existing designs and should not really be used in new designs. That leaves just LS TTL and HC CMOS.

Of these two '74'-series sub-families, LS is slightly faster than HC and is available in a far greater range of functional device types, but generally consumes more supply current/power than HC at frequencies below about 5MHz (*Figure 2.12* compares the performances of 74LS00 and 74HC00 gates). Thus, for most 'new design' applications, the LS TTL and HC CMOS sub-families deserve a joint 'best' award, with a slight edge perhaps going to LS. Note that the rest of this and the next chapter

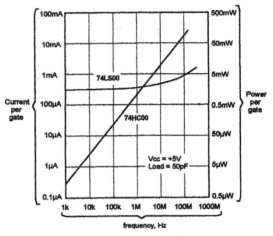

Figure 2.12   Frequency versus current/power graphs of 74LS00 and 74HC00 2-input NAND gates (with a squarewave input)

are concerned mainly with modern TTL devices, and that CMOS devices are next dealt with in Chapter 4.

TTL Logic Levels and Noise Immunity

All digital ICs handle input and output signals that switch between the 'high' (logic-1) or 'low' (logic-0) states. In TTL, each of these 'logic levels' must fall within a defined range of voltage limits. *Figure 2.13* shows the typical input-to-output voltage curve of a 'Standard' TTL inverter that operates from a +5V supply and has a lightly loaded output. Note that the output is 'high', at +3.5V, until the input rises to 0.7V, and then falls fairly linearly as the input is further increased, and eventually stabilizes at a 'low' value of about 0.25V when the input rises above 1.5V. In practice, all Standard and LS TTL ICs are, when using a +5V supply, guaranteed to recognize any input voltage of up to 0.8V as being a logic-0 input, and of 2.0V or above as being a logic-1 input; note the area between these two levels is known as the IC's 'indeterminate' zone or region, and operation within this zone should be avoided.

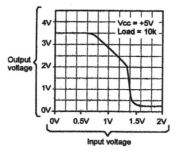

Figure 2.13   Standard-TTL input-to-output voltage graph

In TTL circuitry, different logic levels are used to define input and output signals, since TTL output voltage levels vary considerably with loading conditions; *Figure 2.14* shows how – when an input of 0.4V is applied to the above TTL inverter – the logic-1 output voltage falls from +3.5V at near-zero load current, to a mere 2.0V at a load current of 13mA, and so on. In practice, all Standard TTL ICs are guaranteed (when using a +5V supply) to recognize any output voltage of up to 0.4V as being a logic-0 output, and of 2.4V or above as being a logic-1 output; on LS TTL ICs these levels are 0.5V for logic-0, and 2.7V for logic-1.

When one TTL output is connected directly to a following TTL input, any excessive 'noise' on the output signal may cause incorrect operation of the following input stage. Thus, taking a worst-case situation, a logic-1 Standard TTL output may be as low as 2.4V, and any superimposed negative-going 'noise' pulse greater than 0.4V will drive the following input below the 2.0V

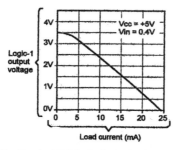

Figure 2.14   Standard-TTL 'logic-1' output voltage–current graph

'logic-0' defined threshold and may cause it to erroneously recognize its input as being a logic-0 (rather than logic-1) signal. The maximum worst-case magnitude of noise that a digital IC can ignore under these conditions is known as its 'noise immunity' or 'noise margin' value, and equals the difference between the logic-0 or logic-1 output/input threshold values. With Standard TTL, noise margins for both logic-1 (NM-H) and logic-0 (NM-L) have defined worst-case values of 400mV; with LS TTL, the noise margins are 700mV for logic-1, and 300mV for logic-0; with CMOS, both margins have values of $V_{DD}/3$. *Figure 2.15* illustrates the values of these three sets of threshold and margin values.

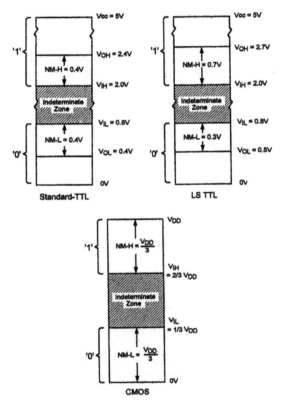

Figure 2.15   Logic level and noise-margin values of Standard TTL, LS TTL, and CMOS

Figure 2.16 expands the above information and shows actual defined threshold voltage and noise margin values, together with typical progagation and power dissipation values for single '00'-type 2-input NAND gates, for the seven major sub-families of TTL (FAST TTL is regarded here as simply a minor variation of AS TTL).

PARAMETER	'74'-Series TTL Sub-families							UNITS
	Standard	L	H	S	LS	AS	ALS	
Propagation Delay (2-input NAND gate)	9nS	33	6	3	8	2	4	ns
Power Dissipation (per gate)	10mW	1	22	20	2	22	1	mW
V_{IH}	2.0V	2.0	2.0	2.0	2.0	2.0	2.0	V
V_{OH}	2.4V	2.4	2.4	2.7	2.7	Vcc-2V	Vcc-2V	V
NM-H	400mV	400	400	700	700	700	700	mV
V_{IL}	0.8V	0.7	0.8	0.8	0.8	0.8	0.8	V
V_{OL}	0.4V	0.3	0.4	0.5	0.5	0.5	0.5	V
NM-L	400mV	300	400	300	300	300	300	mV

Figure 2.16 Typical propagation delay and power dissipation figures for single '00'-type NAND gates within the TTL sub-family ranges, together with sub-family voltage threshold and noise-margin values

Fan-in and Fan-out

In TTL circuitry, an element's input drive requirements are known as its 'fan-in' values, and its output driving capability limits are known as its 'fan-out' values. *Figure 2.17* illustrates the meanings and worst-case values of these items when applied to a Standard TTL element. Thus, *(a)* shows that when the TTL element is driven from a Standard TTL output stage, it draws a worst-case input current (I_{IH}) of 40μA when fed with a 2.4V logic-1 input, but – as shown in *(b)* – feeds 1.6mA (I_{IL}) into the driver when it provides a 0.4V logic-0 input. Diagram *(c)* shows that the TTL element's output can, when in the logic-1 state, provide up to 400μA (I_{OH}) before its output voltage falls below 2.4V; it is thus capable of feeding up to ten Standard inputs, and is said to have a logic-1 'fan-out' (= I_{OH}/I_{IH}) of 10. Similarly, *(d)* shows that the output stage can, when in

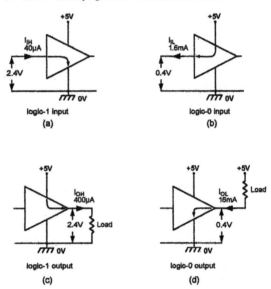

Figure 2.17   Basic input and output parameters of a Standard TTL logic element

the logic-0 state, absorb up to 16mA before its output voltage falls below 0.4V; it is thus capable of driving up to ten Standard inputs, and is said to have a logic-0 fan-out (= I_{OL}/I_{IL}) of 10. Thus, the element has a worst-case fan-out of 10, and it can be used to directly drive as many as ten Standard inputs.

Figure 2.18 presents the above data in tabular form, together with similar data for all other major TTL sub-

PARAMETER	'74'-Series TTL Sub-families							UNITS
	Standard	L	H	S	LS	AS	ALS	
I_{OH}	400µA	200	500	1000	400	2000	400	µA
I_{IH}	40µA	10	50	50	20	20	20	µA
Fan-out, H	10	20	10	20	20	100	20	—
I_{OL}	16mA	2.0	20	20	8.0	20	8.0	mA
I_{IL}	1.6mA	0.18	2.0	2.0	0.36	0.5	0.2	mA
Fan-out, L	10	11	10	10	22	40	40	—
Worst-case fan-out	10	11	10	10	20	40	20	—

Figure 2.18   Fan-in and fan-out values of the major TTL sub-families

families. When working within any one sub-family, note that the most important figure here is the 'worst-case fan-out (F-O)' value. Thus, if you are designing a system based entirely on LS ICs, you can confidently connect an ordinary output directly to as many as 20 normal inputs, without risk of a malfunction due to overloading (if you need to drive more than 20 inputs, you can do so via one or more high-fan-out buffers, etc.). Note that, *within any given sub-family*, all ordinary inputs are said (in TTL jargon) to have a fan-in of unity (1), but that in practice some MSI or LSI ICs (such as counters and registers, etc.) may have special inputs (such as Reset or Preset, etc.) with fan-in values of 2 or greater.

Sometimes, an engineer may have to mix TTL sub-families, usually so that an obsolete IC can be replaced by a readily available modern plug-in close-equivalent. In such a case, it is necessary to relate the fan-out data of one sub-family to that of another, to check that the mix can be made without causing an input or output overload. One easy way of doing this is simply to transpose the data of *Figure 2.18* into 'Standard TTL' fan-in units, as shown in *Figure 2.19*, to gain an approximate idea of the relative 'fan' values of various sub-families. Thus, it can be seen at a glance that LS TTL has only half of the fan-in requirement of Standard TTL, but also has only half of its fan-out capability, etc.

PARAMETER	'74'-Series TTL Sub-families						
	Standard	L	H	S	LS	AS	ALS
Fan-in, '1'	1	0.25	1.25	1.25	0.5	0.5	0.5
Fan-in, '0'	1	0.1125	1.25	1.25	0.225	0.3125	0.125
Fan-in, worst-case	1	0.25	1.25	1.25	0.5	0.5	0.5
Fan-out, '1'	10	5	12.5	25	10	50	10
Fan-out, '0'	10	1.25	12.5	12.5	5	12.5	5
Fan-out, worst-case	10	1.25	12.5	12.5	5	12.5	5

Notes:-

Fan-in, '1' $= \frac{I_{IH}}{40}$ µA

Fan-out, '1' $= \frac{I_{OH}}{40}$ µA

Fan-in, '0' $= \frac{I_{IL}}{1.6}$ mA

Fan-out, '0' $= \frac{I_{OL}}{1.6}$ mA

Fan-in, worst-case = highest figure.

Fan-out, worst-case = lowest figure.

Figure 2.19 TTL fan-in and fan-out in terms of 'Standard TTL' units

An even more useful way of using the basic data of *Figure 2.18* is to convert it into an easily used form that relates the fan-in and fan-out data of each TTL sub-family to all other TTL sub-families, as shown in *Figure 2.20*. Here, by reading across the left-hand columns, it

Sub-family drivers	Sub-family inputs						
	Standard	L	H	S	LS	AS	ALS
Standard TTL	10	40	8	8	20	20	20
L	1.25	11	1	1	5.5	4	10
H	12.5	80	10	10	25	25	25
S	12.5	100	10	10	50	40	50
LS	5	40	4	4	20	16	20
AS	12.5	111	10	10	55	40	100
ALS	5	40	4	4	20	16	20

Note:-

Fan-out = lowest figure of:- $\dfrac{I_{OL}\text{ (driver)}}{I_{IL}\text{ (inputs)}}$ and $\dfrac{I_{OH}\text{ (driver)}}{I_{IH}\text{ (inputs)}}$

Figure 2.20   Maximum number of TTL inputs that may be driven from any TTL sub-family output

can (for example) be seen that a normal LS output can drive up to 5 Standard TTL inputs, and that a Standard TTL output can safely drive up to 20 LS inputs. Thus, if an engineer is faced with a problem such as that illustrated in *Figure 2.21*, in which a fault on an old

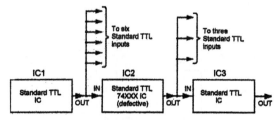

Figure 2.21   IC2 is defective; it is a Standard TTL device. Can it be replaced directly by a 74LSXXX IC? *Figure 2.20* shows that the answer is Yes

Standard TTL circuit is traced to a defective 74XXX-type IC (IC2) which is used to directly drive four other Standard TTL inputs, it can be quickly seen that a 74LSXXX plug-in equivalent IC can be safely used to directly replace the IC2 Standard TTL device without incurring overload problems.

TTL Basic Usage Rules

It is usually a fairly easy matter to design logic circuitry using TTL ICs, providing that a set of TTL basic usage rules are observed. Assuming that the matter of fan-in and fan-out has already been taken care of, there are four 'basic usage' themes outstanding, and these are described in Chapter 3 under the general headings of Power supplies, Input signals, Unused inputs, and Interfacing.

3 TLL Basic Usage Rules

Chapter 2 looked at the '74'-series of ICs and at various sub-members of its TTL and CMOS families, and explained matters such as TTL logic levels, noise immunity, and fan-in and fan-out. The present chapter continues the 'TTL' theme by describing TTL's basic usage rules.

TTL Basic Usage Rules

It is usually a fairly simple matter to design logic circuitry using '74'-series TTL ICs, provided that a set of TTL basic usage rules are observed. Assuming that the matter of fan-in and fan-out has already been taken care of as described in Chapter 2, four other 'basic usage' themes remain, and these are described in the next few pages under the general headings of Power supplies, Input signals, Unused inputs, and Interfacing.

Power Supplies

'74'-series TTL ICs are designed to be used over a very limited supply voltage range (4.75V to 5.25V), and – because they generate very fast pulse edges and have relatively low noise-margin values – must be used with supplies with very low output impedance values (typically less than 0.1 ohms). Consequently, practical TTL circuits should always be powered from a low-impedance well-regulated supply such as one of those shown in *Figures 3.1* to *3.3*, and must be used with a PCB that is very carefully designed to give excellent high-frequency supply decoupling to each TTL IC. In general, the PCB's +5V and 0V supply rail tracks must be as wide as possible (ideally, the '0V' track should take the form of a ground plane), all connections and interconnections should be as short and direct as

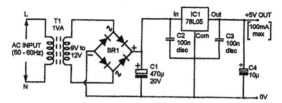

Figure 3.1   5V regulated DC supply (100mA maximum output)

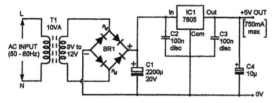

Figure 3.2   5V regulated DC supply (750mA maximum output)

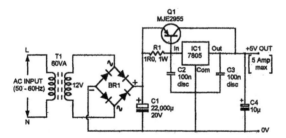

Figure 3.3   5V regulated DC supply (5A maximum output)

possible, the PCB's supply rails should be liberally sprinkled with 4.7µF Tantalum electrolytic capacitors (at least one per 10 ICs) to enhance l.f. decoupling, and with 10nF disc ceramics (at least one per 4 ICs, fitted as close as possible between an IC's supply pins) to enhance h.f. decoupling.

Input Signals

When using TTL, all IC input signals must – unless the IC is fitted with a Schmitt-type input – have very sharp rising and falling edges (typical rise and fall times should be less than 40ns on LS TTL, for example). If rise or fall times are too long, they may allow the input terminal to hover in the TTL element's linear 'indeterminate' zone (see Chapter 2) long enough for the element to burst into wild oscillations and generate spasmodic output signals that may disrupt associated circuitry (such as counters and registers, etc.). If necessary, 'slow' input signals can be converted into 'fast' ones by feeding them to the IC's input terminal via an inverting or non-inverting Schmitt element, as shown in *Figure 3.4*.

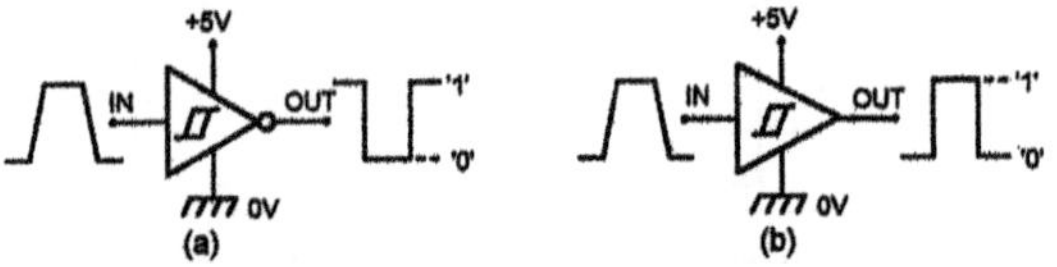

Figure 3.4 'Slow' input signals can be converted into fast ones via *(a)* an inverting or *(b)* non-inverting Schmitt element

Unused Inputs

Unused TTL input terminals should never be allowed simply to 'float', since this makes them susceptible to noise pick-up, etc. Instead, they should be tied to definite logic levels, either by connecting them to V_{CC} via a 1k0 resistor, or shorting them directly to the ground rail, or by connecting them to a TTL input or output terminal that is already in use. *Figure 3.5* shows examples of the four options. The simplest option is to tie the unused input to V_{CC} via a 1k0 resistor, as shown in *(a)*; this resistor has to supply only a few μA of current (I_{IH}) to each input, and can thus easily drive up to 10 unwanted inputs. Alternatively, the input can be tied directly to ground, as in *(b)*, but in this case an input current of several hundred μA (I_{IL}) may flow to the ground rail via the input.

If the unwanted input is on a multi-input gate, it can be disabled by shorting it to one of the gate's used inputs, as in *Figure 3.5(c)*, where a 3-input AND gate is shown

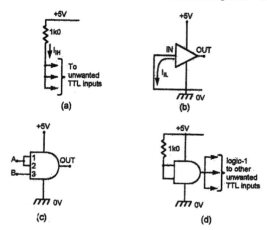

Figure 3.5 Alternative ways of connecting unwanted TTL inputs (see text)

used as a 2-input type. If the IC is a multiple gate type in which an entire gate is unwanted, the gate should be disabled by tying its inputs high if it is a non-inverting (AND or OR) type, or shorting them to ground if it is an inverting (NAND or NOR) type; if desired, the output of this gate can then be used as a fixed logic-1 point that can be used to drive other unwanted inputs, as shown in *(d)*.

Interfacing

An interface circuit is one that enables one type of system to be sensibly connected to a different type of system. In a purely TTL system, in which all ICs are designed to connect directly together, interface circuitry is usually needed only at the system's initial input and final output points, to enable them to merge with the outside world via items such as switches, sensors, relays, and indicators, etc. Occasionally, however, TTL ICs may be used in conjunction with other logic families (such as CMOS), in which case an interface may be needed between the different families. Thus, as far as

TTL is concerned, there are three basic classes of interface circuit, which will now be dealt with under the headings of Input interfacing, Output interfacing, and Logic family interfacing.

Input interfacing

Basically, the digital signals arriving at the inputs of any TTL system must be 'clean' ones with TTL-defined logic-0 and logic-1 levels and with very fast rise and fall times (less than 40ns in LS TTL systems). It is the task of input interfacing circuitry to convert external input signals into this format. *Figures 3.6 to 3.9* show a few simple examples of such circuitry.

Mechanically derived switching signals are notoriously 'bouncy' (see *Figure 1.11*) and must be cleaned up before being fed to a normal TTL input. *Figure 3.6* shows a practical switch-debouncing input interfacing circuit; here, C1 charges – with a time constant of about 10ms – via R1–R2 when S1 is open and generates a logic-0 output via the TTL Schmitt inverter; when S1 is closed it rapidly discharges C1 via R2, driving the Schmitt output high; the effects of any switch-generated 'bounce' signals are eliminated by the circuit's 10ms time constant, and a clean TTL switching waveform is thus available at the Schmitt's output.

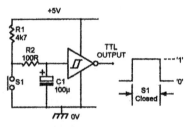

Figure 3.6   Switch-debouncing input interface

Figure 3.7 shows a circuit that can be used to interface almost any clean digital signal to a normal TTL input. Here, when the input signal is below 500mV (Q1's minimum turn-on voltage), Q1 is cut off and the inverting Schmitt TTL output is at logic-0; when the input is significantly above 600mV, Q1 is driven on and the Schmitt output goes to logic-1. Note that the digital input

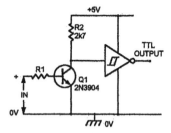

Figure 3.7 Transistor input interface

signal can have any maximum voltage value, and R1 is chosen simply to limit Q1's base current to a safe value.

Figure 3.8 is a simple variation of the above circuit, with the transistor built into an optocoupler; the circuit action is such that the Schmitt's output is at logic-0 when the optocoupler input is zero, and at logic-1 when the input is high; note that the optocoupler provides total electrical isolation between the input and TTL signals.

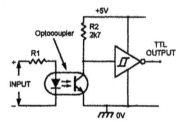

Figure 3.8 Optocoupler input interface

Finally, *Figure 3.9* is another simple circuit variation, with the basic digital input signal fed to Q1's base via the R1–C1–R2–C2 low-pass filter network, which

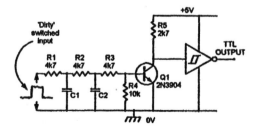

Figure 3.9 'Dirty-switching' input interface

eliminates unwanted high-frequency components and thus can convert very 'dirty' input signals (such as those from vehicle contact-breakers, etc.) into a clean TTL format.

Output interfacing

Most TTL ICs have normal totem-pole output stages, but some of them have modified totem-pole outputs with 3-state (Tri-State) gating; a few TTL ICs have open-collector (o.c.) totem-pole output stages. Note that normal totem-pole outputs should not (except in a few special cases) be connected in parallel. TTL o.c. outputs *can* be connected in parallel, however, and 3-state ones can be connected in parallel under special conditions; basic ways of using o.c. and 3-state outputs are described in later chapters.

A normal totem-pole output stage can source or sink useful amounts of output current, and can be used in a variety of ways to interface with the outside world. A few simple examples of such circuits are shown in *Figures 3.10* to *3.17. Figure 3.10* shows a couple of ways of driving LED output indicators via non-inverting TTL elements. Note that a normal TTL output can sink fairly high load currents (typically up to 50mA in an LS device), but has an internally limited output sourcing ability; thus, the LED current must be limited to a safe value via R1 if it is connected as in *(a)*, but is internally limited in *(b)*. *Figure 3.11* shows alternative ways of driving LEDs, using inverting TTL elements.

Figure 3.12 shows two current-boosting load-driving output interface circuits, in which the load uses the same power supply as the TTL circuit. In *(a)*, npn transistor

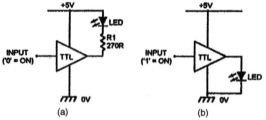

Figure 3.10   LED-driving output interfaces, using non-inverting TTL elements

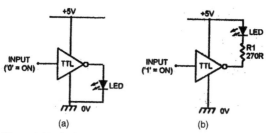

Figure 3.11 LED-driving output interfaces, using inverting TTL elements

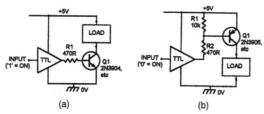

Figure 3.12 Current-boosting load-driving output interfaces

Q1 is cut off when the input of the non-inverting TTL element is at logic-0, and is driven on via R1 when the input is at logic-1. The reverse action is obtained in *(b)*, where pnp transistor Q1 is pulled on via R2 when the input is at logic-0, and is cut off via pull-up resistor R1 when the input is at logic-1.

Figure 3.13 shows two output interface circuits that can be used to drive loads that use independent positive supply rails. Q1 is turned on by a logic-1 input in *(a)*, and a logic-0 input in *(b)*. If the external load is inductive (such as a relay or motor, etc.), the circuits should be fitted with protection diodes, as shown dotted in the diagrams.

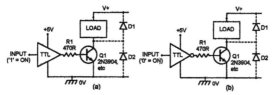

Figure 3.13 Output interface to load with independent positive rail

Figure 3.14 shows two optocoupled output-interface circuits that can be used to drive loads that use fully independent DC power supplies; the load is turned on via a logic-1 input in *(a)*, and a logic-0 input in *(b)*. Note that the optocoupler input (the LED) could alternatively be connected between the +5V rail and the TTL output via a current-limiting resistor, using the same basic connections as *Figure 3.10(a)* or *3.11(b)*.

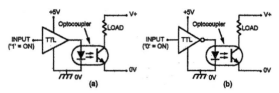

Figure 3.14 Optocoupled output interface

Figure 3.15 shows an output interface that can be used to control a low-power lamp or similar resistive load that is driven from AC power lines and consumes no more than about 100mA of current. This circuit uses an optocoupled triac, and these typically need an LED input current of less than 15mA and can handle triac

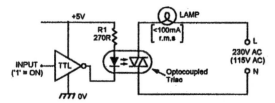

Figure 3.15 Output interface to a low-power AC lamp via an optocoupled triac

load currents of up to about 100mA mean (500mA surge) at up to 400V peak. Note that optocoupled triacs are best used to activate a high-power 'slave' triac, that can drive a load of any desired power rating. *Figures 3.16* and *3.17* show two such circuits.

The *Figure 3.16* circuit is suitable for use with non-inductive loads such as lamps and heating elements. It can be modified for use with inductive loads such as motors by using the connections of *Figure 3.17*, in

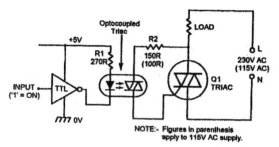

Figure 3.16 Output interface to a high-power non-inductive AC load

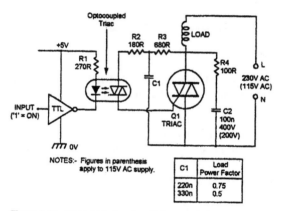

Figure 3.17 Output interface to a high-power inductive load

which R2–C1–R3 provide a degree of phase-shift to the triac gate-drive network, to ensure correct triac triggering action, and R4–C2 form a snubber network, to suppress rate effects.

Logic family interfacing
It is generally bad practice to mix different logic families in any system, but on those occasions where it does occur the mix is usually made between TTL and CMOS devices that share a common 5 volt power supply; in this case the form or necessity of any interfacing circuitry depends on the direction of the interface and on the precise sub-families that are involved; *Figures 3.18* to *3.21* show the four most useful types of interface arrangement.

The output of any TTL element can be used to drive any normal CMOS logic IC (including some sub-members of the '74'-series) by using the connections shown in *Figure 3.18*, in which R1 is used as a TTL pull-up resistor and ensures that the CMOS consumes minimal quiescent current when the TTL output is in the logic-1 state.

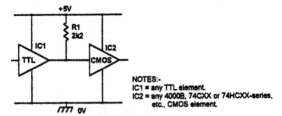

Figure 3.18   TTL-to-CMOS interface

Standard 4000B-series and 74CXX-series CMOS elements have very low fan-outs, and can only drive a single Standard TTL or LS TTL element, as shown in *Figures 3.19* and *3.20*. 74HCXX-series (and 74ACXX-series) CMOS elements, on the other hand, have excellent fan-outs, and can directly drive up to 2 Standard TTL inputs, or 10 LS TTL inputs, or 20 ALS TTL inputs, as shown in *Figure 3.21*.

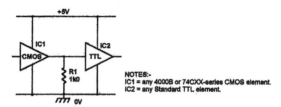

Figure 3.19   CMOS-to-Standard-TTL interface

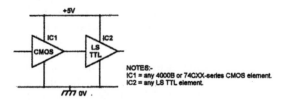

Figure 3.20   CMOS-to-LS-TTL interface

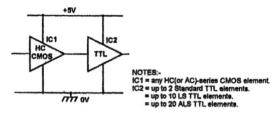

Figure 3.21 HC-CMOS-to-TTL interface

In cases where the TTL and CMOS ICs use individual positive supply rails (5V for TTL, 3V to 18V for CMOS), an interface can be made between the two systems by using a direct-coupled npn transistor as a level-shifter between them, as shown in *Figures 3.22* and *3.23* (these simple circuits may need some refining if they are to be used at frequencies above a few hundred kHz). Finally, note that if the TTL element has an o.c. totem-pole output, a direct interface can sometimes be made between the TTL output and the input of an individually powered CMOS element, etc.; the basics of this technique are described in Chapter 4.

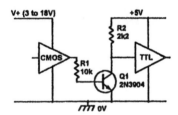

Figure 3.22 CMOS-to-TTL interface, using independent positive supply rails

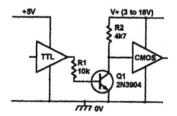

Figure 3.23 TTL-to-CMOS interface, using independent positive supply rails

4 Modern CMOS Digital ICs

One of the most important events in the history of digital electronics was the development, in 1969, of the new IC technology known as CMOS. CMOS (*Complementary-symmetry MOS*-FET) digital IC elements have major advantages over TTL types. They are simple and inexpensive, consume near-zero quiescent current, have a very high input impedance, can operate over a wide range of supply voltages, have excellent noise immunity, and are very easy to use. In 1972, practical CMOS arrived on the commercial scene in the form of a brand-new medium-speed family of digital ICs known as the '4000'-series. This new family was not as fast as the TTL technology then in use in the rival '74'-series of digital ICs, but in the mid 1980s a new high-speed type of CMOS was developed and introduced as a new member of the '74' family of devices. The advantages of this new 'fast' CMOS were so great that in 1994 it overtook TTL in popularity within the '74'-series, finally making CMOS *the* most popular of all modern digital IC technologies. This chapter explains the operating principles of these '4000'- and '74'-series CMOS devices, and describes CMOS basic usage rules.

CMOS Basics

The most basic element in any digital IC family is the digital inverter. *Figure 4.1* (repeated from Chapter 1) shows a basic CMOS inverter. It is a 'totem-pole' type of amplifier and consists of a complementary pair of enhancement-mode MOSFETs wired in series between the two supply lines, with p-channel MOSFET Q1 at the top and n-channel MOSFET Q2 below and with the MOSFET gates (which have a near-infinite DC input impedance) tied together at the input terminal and the output taken from the junction of the two devices. The pair can be powered from any supply in the 3V to 15V range. The basic digital action of the n-channel device is such that its drain-to-source path acts like an open-

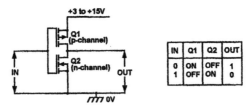

Figure 4.1 Circuit and Truth Table of a basic CMOS inverter

circuit switch when the input is at logic-0, or as a closed switch in series with a 400R resistor when the input is at logic-1. The p-channel MOSFET has the inverse of these characteristics, and acts like a closed switch plus a 400R resistance with a logic-0 input, and an open switch with a logic-1 input. The basic action of the CMOS inverter can be understood with the help of *Figure 4.2.*

Figure 4.2(a) shows the digital equivalent of the CMOS inverter circuit with a logic-0 input. Under this condition, Q1 (the p-channel MOSFET) acts like a closed switch in series with 400R, and Q2 acts like an open switch. The circuit thus draws zero quiescent current but can 'source' fairly large drive currents into an external output-to-ground load via the 400R output resistance (R1) of the inverter. *Figure 4.2(b)* shows the inverter's equivalent circuit with a logic-1 input. In this case Q1 acts like an open switch, but Q2 (the n-channel MOSFET) acts like a closed switch in series with 400R; the inverter thus draws zero quiescent current under this condition, but can 'sink' fairly large currents from an external supply-to-output load via its internal 400R output resistance (R2).

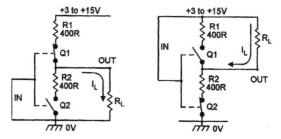

Figure 4.2 Equivalent circuits of the CMOS digital inverter with *(a)* logic-0 and *(b)* logic-1 inputs

Thus, the basic CMOS digital inverter can be used with any supply in the 3V to 15V range, has a near-infinite input impedance, draws near-zero (typically 0.01μA) supply current with a logic-0 or logic-1 input, can source or sink substantial output currents, and has an output impedance of about 400R. Note that, unlike the TTL inverter, its output can swing all the way from zero to the full positive supply rail value, since no potentials are lost via saturation or forward biased junction voltages, etc. Typically, a basic (mid 1970s style) CMOS stage has a propagation delay ranging from 12ns when using a 12V supply, to 60ns at 3V, etc.

The '4000A'-series of ICs

The initial 1972 range of digital ICs was known as the '4000A'-series; it used the basic type of CMOS inverter shown in *Figure 4.1*, but incorporated extensive diode–resistor 'clamping' networks to protect its MOSFETs against damage from static charges, etc. Thus, a complete 'A'-series inverter stage took the basic form shown in *Figure 4.3*.

Commercial testing of the early 'A'-series range of CMOS devices quickly revealed a number of design problems. Their on–off resistance values were, for example, very sensitive to gamma radiation effects, thus limiting their value in outer-space projects, and they gave uneven 'high' and 'low' output impedances and propagation delays, i.e. they had poor output symmetry.

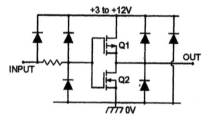

Figure 4.3 Basic '4000A'-series inverter stage, with internal input and output protection networks

But the most important problem was that their output switching levels were overly sensitive to the magnitudes of their input switching signals; the root cause of this problem can be understood with the aid of *Figure 4.4*, which shows the linear characteristics of the CMOS inverter's two MOSFETs when they are operated from a 15 volt supply.

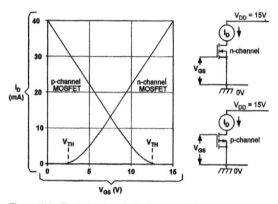

Figure 4.4 Typical gate-volts/drain-current characteristics of p- and n-channel MOSFETs operated from a 15 volt supply

Note in *Figure 4.4* that each MOSFET acts like a voltage-controlled resistance. The n-channel device has a near-infinite drain-to-source resistance at zero input voltage: the resistance remains high until the input rises to a 'threshold' value of about 1.5 to 2.5 volts, but then decreases as the input voltage is increased, eventually falling to about 400R when the input equals the supply line voltage. The p-channel MOSFET has the reverse of these characteristics. Thus, when the two MOSFETs are wired in series and used as a 15 volt basic CMOS inverter they produce the typical drain-current transfer graph shown in *Figure 4.5*, and the voltage transfer graph of *Figure 4.6*; these graphs can be explained as follows.

Suppose in *Figures 4.5* and *4.6* that the CMOS inverter's input voltage is slowly increased upwards from zero. The inverter current is near zero until the input exceeds the n-channel MOSFET's threshold voltage, at which point its resistance starts to fall and

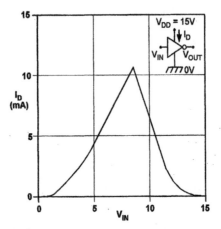

Figure 4.5   Typical drain-current transfer characteristics of the simple CMOS inverter

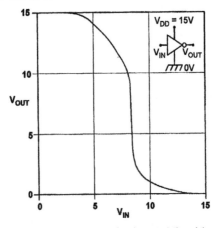

Figure 4.6   Typical voltage transfer characteristics of the simple CMOS inverter

that of the p-channel MOSFET starts to increase. Under this condition the inverter current is dictated by the larger of the two resistances; when the input is far less than half-supply volts, the n-channel MOSFET resistance is far greater than that of the p-channel device, so the output is high (at logic-1). When the input is at a *transition* value somewhere between 30 and 70% of the supply voltage the two MOSFETs have similar resistance values and the inverter acts as a linear amplifier with a voltage gain of about 30dB and draws several milliamps of supply current; under this condition, small changes of input voltage cause large changes of output voltage. When the input is further increased, well above half-supply volts, the resistance of the n-channel MOSFET falls below that of the p-channel device, and the output goes low (to logic-0). Finally, when the input rises above the threshold value of the p-channel MOSFET it acts like an open switch, and the inverter current again falls to near zero.

Thus, the 'A'-series type of inverter gives an output that switches fully between the supply rail values only if its input voltage swings well above and below its two internal 'threshold' voltage values. Note (from *Figure 4.5*) that CMOS draws a brief pulse of supply current each time it goes through a switching transition; the more often CMOS changes state in a given time, the greater are the number of current pulses that it takes from the supply and the greater is its *mean* current consumption. Thus, CMOS current consumption is directly proportional to switching frequency.

The '4000B'-series of ICs

The defects of the '4000A'-series were so severe that an improved CMOS series, known as the '4000B'- or 'buffered' series, was introduced in about 1975, and the old '4000A'-series was slowly phased out of production. The major feature of this new series is that each of its 'inverters' consists of three basic inverters wired in series, as shown in *Figure 4.7*, so that each 'buffered'

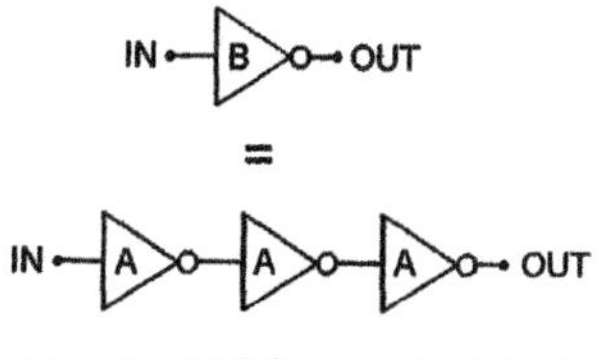

Figure 4.7 A 'B'-series CMOS inverter can be made by wiring three A-series types in series

inverter has a typical linear voltage gain of 70 to 90dB and has the typical voltage transfer graph of *Figure 4.8*, in which any input below $V_{DD}/3$ is recognized as a logic-0 input and any input above $2V_{DD}/3$ is recognized as a logic-1 input. Other changes in the new series include greatly improved output-drive symmetry and immunity to gamma-radiation effects, new and better input and output protection networks (see *Figure 4.9*), and improved voltage ratings (usually to 15V maximum, but to 18V maximum in some manufacturer's versions, compared to 12V maximum in the original 'A'-series).

One disadvantage of the 'B'-series is that its propagation delays are larger than those of the old 'A'-series. To counter this problem, a few new-generation

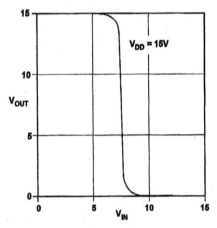

Figure 4.8   Voltage transfer graph of the *Figure 4.7* 'B'-series inverter

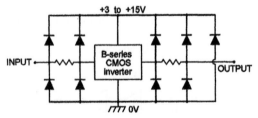

Figure 4.9   Basic '4000B'-series inverter with typical input and output protection networks

devices are produced in an 'unbuffered' format (denoted by a 'UB' suffix), but incorporate all the other improvements of the 'B'-series. Typically, 'UB' inverters have an AC gain of 23dB at 10 volts, and are useful in several analogue applications (see Chapter 5). Note that the bandwidth and propagation delays of a CMOS device vary with supply voltage and with capacitive output loading; *Figure 4.10* lists the typical propagation delays of both 'UB' and 'B'-series inverters when used with supply values of 5V, 10V, and 15V when driving a 50pF load.

V_{DD}	Typical Propagation Delays	
	UB-types	B-series types
15V	20nS	35nS
10V	25nS	50nS
5V	40nS	120nS

Figure 4.10 Typical propagation delays of '4000B'-series inverters when driving a 50pF load

The '4500B'-series of ICs

The '4000B'-series range of ICs consists mainly of fairly simple SSI or MSI devices such as logic gates and simple counters, etc. In the late 1970s and early 1980s a number of more complex MSI and LSI 'B'-type CMOS ICs such as encoders, decoders, and presettable counters (etc.) were introduced; these advanced devices carry '45XX' or '47XX' numbers, and are generally known as the '4500'-series of CMOS ICs.

'Fast' CMOS ICs

In the early 1980s, engineers strove to design a really fast type of CMOS that could outclass 'LS' TTL when operated from a 5 volt supply and could thus become *the* dominant technology within the '74'-series of ICs. Normal CMOS is based on MOSFET (Metal-Oxide Silicon FET) technology, and this is simply a variation of IGFET (Insulated-Gate FET) technology. Specifically, a MOSFET device is an IGFET device that uses metal-oxide gate insulation, and the first big step in developing 'fast' CMOS was to use silicon-oxide rather than metal-oxide gate insulation in the basic IGFETs. This simple measure resulted in a dramatic reduction in the IGFET's internal input capacitance and an equally dramatic increase in operating speed. The next step was to apply these new IGFETs to the basic CMOS configuration, and when this was done and significant changes were made in the element's geometry, the resulting device acted like normal CMOS but was as fast as 'LS' TTL when operated from a 5 volt supply and (unlike some other versions of CMOS) had excellent output drive capability. Strictly, this new device should have been given a special name such as CSOS (Complementary Silicon-Oxide Silicon FET), but instead was simply christened 'fast' CMOS.

'Fast' CMOS has many similarities with conventional '4000B'-series CMOS. It is available in both buffered (triple-inverter) and unbuffered (single-inverter) basic versions, and has all inputs and outputs protected via internal diode–resistor networks. It can (in most cases) use any supply in the 2V to 6V range, and when first introduced was intended to replace many existing devices in the '74'-series of ICs. Since then, however, it has also been used to make 'fast' versions of many popular devices within the '4000B'- and '4500B'-series of ICs.

CMOS '74'-series sub-families

When the '74'-series of ICs first appeared in 1972 it was based entirely on TTL technology, which inherently consumes a fairly high quiescent current. In the late 1970s, a slightly modified version of standard CMOS (optimized for 5V operation) was introduced as a new 'C' sub-family within the '74'-series range of devices, and offered the advantage of near-zero quiescent current consumption. This 'C' sub-family was too slow and had too weak an output-drive capability to obtain great popularity, but in later years the 'fast' type of CMOS was developed specifically for use in the '74'-series, as already described, and so far a total of five CMOS sub-families has been introduced in the '74'-series, as follows.

Standard (C) CMOS (now obsolete). This was virtually normal MOSFET-type CMOS in a '74'-series format. Typically, a single 74C00 2-input NAND gate consumed about 15mW at 10MHz, and had a propagation delay of 60ns at 5V.

High-speed (HC) CMOS. Introduced in the early 1980s, this is the basic 'fast' silicon-oxide version of CMOS, and gives speed performances similar to LS TTL, but with CMOS levels of power consumption. HC '74'-series devices using this technology have CMOS-compatible inputs; typically, a single 74HC00 2-input NAND gate consumes less than 1μA of quiescent current, and has a propagation delay of 8ns at 5V.

High-speed (HCT) CMOS. These are fast HC-type devices, but have TTL-compatible inputs and are meant to be driven directly from TTL outputs. Typically, a 74HCT00 2-input NAND gate consumes less than 1μA of quiescent current and has a propagation delay of 18ns.

Advanced high-speed (AC) CMOS. In the late 1980s, further advances in high-speed CMOS design and fabrication techniques yielded even better speed performances. AC '74'-series devices using this technology have CMOS-compatible inputs; typically, a 74AC00 2-input NAND gate has a propagation delay of 5ns.

Advanced high-speed (ACT) CMOS. These are AC-type devices, but have TTL-compatible inputs and are meant to be driven from TTL outputs. Typically, a 74ACT00 2-input NAND gate has a propagation delay of 7ns.

Basic CMOS Circuit Variations

There are three important variations of the basic CMOS circuit that are often used in ICs in the medium-speed '4000B'-series and fast '74'-series ranges of devices. The first of these is the 'open-drain' configuration, which is used in some inverters and buffers, etc. *Figure 4.11* shows a typical open-drain inverter, which is configured like a normal high-gain 3-stage CMOS inverter except that the final stage consists of a single n-channel enhancement-mode IGFET (Q1) that has its drain connected directly to the circuit's output terminal. The circuit's action is such that Q1 is cut off when the input is at logic-0, and is driven on when the input is at logic-1. The circuit can be used to directly drive an external load that is connected between 'OUT' and the +ve supply rail, in which case the load activates when a logic-1 input is applied.

The second variation concerns the use of a '3-state' type of output that in normal use gives a conventional logic-0 or logic-1 low-impedance output, but can also be set to a third state in which the output is effectively open-circuit. This facility is useful in allowing several outputs or inputs to be wired to a common bus and to communicate along that bus by ENABLING only one output and one input device at a time. *Figure 4.12* shows the typical circuit of a non-inverting buffer of this type, together with its Truth Table. Thus, when the DISABLE INPUT control is at logic-0, the circuit gives normal 'buffer' operation; under this condition Q1 is driven OFF and Q2 is driven ON when IN is at logic-0, thus driving OUT to logic-0; the reverse of this action is obtained when the input is at logic-1. When the DISABLE INPUT control

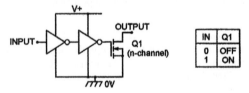

Figure 4.11   Basic CMOS inverter with open-drain (o.d.) output

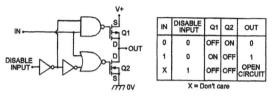

IN	DISABLE INPUT	Q1	Q2	OUT
0	0	OFF	ON	0
1	0	ON	OFF	1
X	1	OFF	OFF	OPEN CIRCUIT

X = Don't care

Figure 4.12 Basic circuit of a 3-state CMOS non-inverting buffer

is set to logic-1, both Q1 and Q2 are driven OFF, irrespective of the state of the IN input, and under this condition OUT is effectively disabled, and acts as an open circuit. *Figure 4.13* shows the simplified equivalent circuit of this buffer when it is in its high-impedance output state.

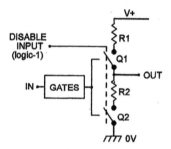

Figure 4.13 Equivalent of a 3-state buffer circuit in its 'third' high-impedance state

The third CMOS circuit variation is that of the 'bilateral switch' or transmission gate. The basic action of any enhancement-mode IGFET is such that its drain-to-source path acts like a near-perfect unidirectional switch: when the IGFET is OFF, the path acts like an open circuit, and when it is ON it acts like a low-value resistor and (unlike a bipolar transistor) does not suffer from saturation-voltage problems, etc. When turned on, an n-channel IGFET passes current from drain-to-source, and a p-channel IGFET passes current from source-to-drain. Thus, a near-perfect bidirectional or 'bilateral' electronic switch can be made by wiring an n-channel and a p-channel IGFET in parallel (source-to-source and drain-to-drain) and driving their gates in anti-phase, as shown in *Figure 4.14*. Here, both IGFET paths are effectively open when the CONTROL input is

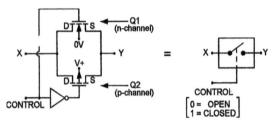

Figure 4.14 Basic CMOS bilateral switch or 'transmission gate'

at logic-0, and closed when the CONTROL input is at logic-1. Under the 'closed' condition, current can flow from X to Y via Q1, or from Y to X via Q2; current can thus flow in either direction between these points, and the circuit thus simulates a simple electro-mechanical switch.

CMOS Basic Usage Rules

CMOS ICs are very easy to use. They are very tolerant of supply voltage variations and, unlike TTL types, present very few input-drive/output-drive matching problems. There are, in fact, only seven 'basic usage' themes to consider when dealing with CMOS, and these will now be dealt with under the headings of Type selection, Handling CMOS, Power supplies, Input signals, Unused inputs, and Interfacing.

Type selection
The question 'which CMOS family should I use?' can easily be answered with the help of *Figure 4.15*, which lists the major characteristics of the six readily-available modern CMOS sub-families and compares them with those of LS TTL. Of these types, the 4000UB sub-family is only available in the form of a few simple buffer and inverter ICs, and should be regarded as a simple variant of the main 4000B sub-family, and the 74HCT and 74ACT types are meant to be directly driven from TTL outputs, and are of use only in a few specialized applications.

	LS TTL	4000B CMOS	4000UB CMOS	74HC CMOS	74HCT CMOS	74AC CMOS	74ACT CMOS
SUPPLY VOLTAGE RANGE	4.75 - 5.25V	3 - 15V	3 - 15V	2 - 6V	4.5 - 5.5V	2 - 6V	4.5 - 5.5V
QUIESCENT CURRENT (PER GATE)	0.5mA	.01µA	.01µA	.02µA	.02µA	.02µA	.02µA
PROPAGATION DELAY (PER GATE)	9nS	125nS @ 5V 50nS @ 10V 40nS @ 15V	90nS @ 5V 50nS @ 10V 40nS @ 15V	8nS @ 5V	10nS @ 5V	5nS @ 5V	7nS @ 5V
MAXIMUM OPERATING FREQUENCY (COUNTER)	40MHz @ 5V	2MHz @ 5V 5MHz @ 10V 6MHz @ 15V	—	40MHz @ 5V	—	100MHz @ 5V	—
FAN-OUT [@ 5V, to LS TTL INPUTS]	20	1	1	10	10	60	60

Figure 4.15   Table showing general characteristics of LS TTL and the six major CMOS digital IC types

Of the remaining three CMOS sub-families (4000B, 74HC, and 74AC), the 4000B sub-family can be used in any application that requires the use of a supply in the range 3V to 15V and in which maximum operating frequencies do not exceed 2MHz at 5V, or 6MHz at 15V. Alternatively, if supply voltages are restricted to the 2V to 6V range, the 74HC sub-family can be used to operate at frequencies up to 40MHz at 5V, or the 74AC sub-family at frequencies up to 100MHz at 5V.

Note that all TTL ICs have special input-drive requirements, and the 'fan-out' numbers in *Figure 4.15* show how many parallel-connected standard LS TTL inputs can be directly driven from the output of each listed sub-family member. Thus, 4000B CMOS can only drive one such input, but 74HC and HCT CMOS can each drive 10 such inputs, and 74AC and ACT can each drive up to 60 LS TTL inputs.

Handling CMOS

CMOS is based on high-impedance IGFET technology, which – when being handled – is easily damaged by high-voltage static charges of the type that can build up on the body of the person handling them. All modern CMOS digital ICs incorporate extensive internal diode-clamping circuitry that is designed to protect their internal IGFETs against damage from 'reasonable' values of static discharge of this type when the IC is being handled.

Figure 4.16(a) shows the basic laboratory circuit that is used – when testing CMOS ICs – to simulate 'reasonable' values of static discharge from a human body; C1 has a value of 100pF and simulates the typical body capacitance of a charged human adult, and R1 has a value of 1k5 and simulates the body's typical discharge resistance. When a CMOS IC is being giving evaluation tests, C1 is charged to a high-value test voltage via S1, and is then applied to two of the IC's test points via S1 and R1. A basic CMOS element has four

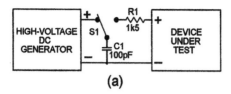

(a)

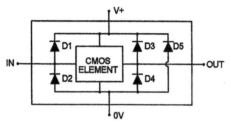

Figure 4.16 *(a)* Typical 'electrostatic discharge test circuit', and *(b)* simple equivalent of a CMOS digital IC element

terminals (IN, OUT, V+, and 0V), and thus has a total of 12 possible 2-pin test permutations, and the test circuit is applied to each of these 2-pin permutations in a full test sequence. Typically, modern CMOS digital ICs are expected to survive a test voltage of 2.5kV in all of these test modes.

Figure 4.16(b) shows the basic form of a CMOS element's internal protection circuitry; here, D1 or D2 conduct if IN tries to go above V+ or below 0V, D3 or D4 conduct if OUT tries to go above V+ or below 0V, and D5 conducts if 0V tries to go above V+; D5 also conducts in the Zener mode if V+ goes more than about 20V above 0V.

It is important to understand the meaning of these CMOS 'static discharge protection' tests. Suppose that a 3kV test voltage is applied between the IC's reverse-connected 0V and V+ pins. Under this condition, D5 is forward biased, and C1 discharges via D5 and R1; R1 limits C1's peak discharge current to 2A and gives it a basic time constant of 150ns. Thus, D5 passes only a very brief spike of forward current as C1 discharges; if D1's thermal time constant is very long compared to the period of the spike, it may not suffer damage from this test, even though it can only handle normal DC currents of (say) 25mA maximum. Note that the peak voltage appearing across D5 in this test is roughly 1V, most of C1's 3kV discharge voltage being lost across R1.

The protection networks used in CMOS ICs are not designed to be effective against massive values of static discharge, such as the several thousand volts that may be generated by a person vigorously prancing about on a nylon carpet. Consequently, when handling 'naked' CMOS ICs, always take sensible precautions against the build-up of large static charges; do not wear nylon clothing or use nylon mats/carpets in the workshop, and make sure that soldering irons, etc., are correctly grounded. To be really safe, wear a grounded metal wrist strap when working with CMOS, particularly when soldering. Note, however, that in reality it is very unlikely that you will *ever* damage a CMOS IC in normal handling, even if you don't bother to wear a grounded wrist strap.

Power supplies
CMOS ICs of the 4000B and 74HC and 74AC types are designed to operate over a wide range of supply voltages, and can thus be powered from batteries or from regulated or unregulated power supplies. 74HCT and 74ACT types, however, are designed to operate from supplies in the 4.5V to 5.5V range, and must be powered from low-impedance well-regulated supplies of the types shown in *Figures 3.1* to *3.3* (in Chapter 3).

All CMOS ICs generate fast pulse-switching edges. Consequently, most CMOS circuits should be used with a PCB that is designed to give excellent high-frequency supply decoupling to each IC. In general, the PCB's supply and ground-rail tracks must be as wide as possible (ideally, the '0V' track should take the form of a ground plane), all connections and interconnections should be as short and direct as possible, the PCB's supply rails should be liberally sprinkled with 4.7µF Tantalum electrolytic capacitors (at least one per 10 ICs) to enhance l.f. decoupling, and with 10nF disc ceramics (at least one per 4 ICs, fitted as close as possible between an IC's supply pins) to enhance h.f. decoupling.

When experimenting with CMOS ICs, never allow the power supply to be connected in the wrong polarity, since this will cause heavy supply currents to flow through the IC's protective diode networks (specifically, through D5 in *Figure 4.16*) and cause instant damage to the IC's substrate.

Input signals

When using CMOS, all IC input signals must – unless the IC is fitted with a Schmitt-type input – have very sharp rising and falling edges. If rise or fall times are too long, they may allow the input terminal to hover in the CMOS element's linear zone long enough for the element to burst into wild oscillations and generate spasmodic output signals that may disrupt associated circuitry (such as counters and registers, etc.). If necessary, 'slow' input signals can be converted into 'fast' ones by feeding them to the IC's input terminal via CMOS Schmitt elements.

One possible way of damaging CMOS is via a *very low-impedance* input or output signal that is either connected to the CMOS when its power supply is switched off, or is of such large amplitude that it forces the input terminal well above the positive supply line or below the zero-volts rail, thus causing a damaging current to flow through one or more of the IC's protection diodes (specifically, through *Figure 4.16*'s input diodes D1 or D2, or output diodes D3 or D4). The possibility of such damage can be eliminated by wiring a 1k0 resistor in series with each input/output terminal, to limit such currents to safe values of a few milliamps.

Unused inputs

Unused CMOS input terminals must never be allowed simply to 'float', but must always be tied to definite logic levels by either connecting them directly to the supply or ground rails (depending on the IC's logic requirements), or to some other point with well-defined logic levels. *Figure 4.17* shows some of the available options. If the unwanted input is on a multi-input gate, it can be disabled by shorting it to one of the gate's used inputs, as in *Figure 4.17(c)*, where a 3-input AND gate is shown used as a 2-input type. If the IC is a multiple

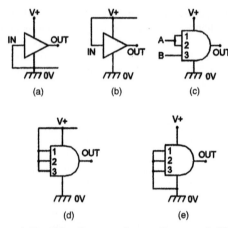

Figure 4.17　Alternative ways of connecting unwanted CMOS inputs (see text)

gate type in which an entire gate is unwanted, the gate should be disabled by tying all of its inputs to a common high or low point, as in *(d)* and *(e)*.

All 'used' CMOS input terminals must also be tied to definite logic levels, and must never be allowed to 'float'. *Figure 4.18* shows three commonly used options. In *(a)*, the input is normally tied low by R1, and in *(b)* it is normally tied high by R1. In *(c)*, the input is direct-coupled to the output of a driving stage, which determines the input logic level.

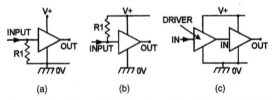

Figure 4.18　All 'used' CMOS inputs must be tied to definite logic levels (see text)

Interfacing

An interface circuit is one that enables one type of system to be sensibly connected to a different type of system. In a purely CMOS system, in which all ICs are designed to connect directly together, interface circuitry is usually needed only at the system's initial input and final output points, to enable them to merge with the outside world via items such as switches, sensors, relays, and indicators, etc. Occasionally, however, CMOS ICs may be used in conjunction with other logic families (such as TTL), in which case an interface may be needed between the different families. Thus, as far as CMOS is concerned, there are three basic classes of interface circuit, which will now be dealt with under the headings of Input interfacing, Output interfacing, and Logic family interfacing.

Input interfacing
The digital signals arriving at the inputs of a CMOS system must be 'clean' ones with well-defined logic levels and with fast rise and fall times. It is the input interfacing circuitry's task to convert external input signals into this format. *Figures 4.19* to *4.22* show four simple examples of such circuitry; these circuits are similar to the TTL designs of *Figures 3.6* to *3.9*, but must use CMOS Schmitt elements and can use any positive supply rail voltage within the operating limits of the CMOS element.

The *Figure 4.19* circuit is designed to clean up the 'dirty' switching signals of push-button switch SW1 and convert them into a form suitable for driving a normal CMOS input. Here, the input of the Schmitt buffer is tied to ground via R1 and R2 and is normally low. When SW1 is closed, C1 rapidly charges up and drives the Schmitt output high, but when SW1 opens again C1 discharges relatively slowly via R1, and the Schmitt output does not return low again until roughly 20ms later. The circuit thus ignores the transient switching

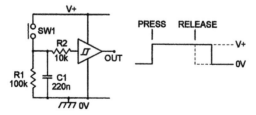

Figure 4.19   CMOS 'noiseless' push-button switch

effects of SW1 noise and contact bounce, etc., and generates a clean output 'switching' waveform with a period that is roughly 20ms longer than the mean duration of the SW1 switch closure.

Figure 4.20 shows a circuit that can be used to interface almost any clean digital signal to a normal CMOS input. Here, when the input signal is below 500mV (Q1's minimum turn-on voltage), Q1 is cut off and the inverting Schmitt's output is at logic-0; when the input is significantly above 600mV, Q1 is driven on and the Schmitt output goes to logic-1. Note that the digital input signal can have any maximum voltage value, and R1 is chosen simply to limit Q1's base current to a safe value.

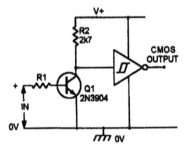

Figure 4.20   CMOS 'transistor input' interface

Figure 4.21 is a simple variation of the above circuit, with the transistor built into an optocoupler; the circuit action is such that the Schmitt's output is at logic-0 when the optocoupler input is zero, and at logic-1 when the input is high; note that the optocoupler provides total electrical isolation between the input and CMOS signals.

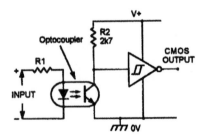

Figure 4.21   CMOS 'optocoupler input' interface

Finally, *Figure 4.22* is another simple circuit variation, with the basic digital input signal fed to Q1's base via the R1–C1–R2–C2 low-pass filter network, which eliminates unwanted high-frequency components and thus can convert very 'dirty' input signals (such as those from vehicle contact-breakers, etc.) into a clean CMOS format.

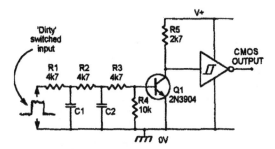

Figure 4.22 CMOS 'dirty-switching' input interface

Output interfacing
CMOS totem-pole output stages are designed to source or sink fairly high peak values of output current. Consequently, if the output is shorted directly to the IC's zero-volts or positive supply rail, the resulting DC output currents can in some cases be so high that the IC may be damaged. Thus, when a CMOS IC is used to drive a DC load, its load current must always be limited to a safe value. *Figure 4.23* shows the typical short-circuit output currents of two different manufacturer's

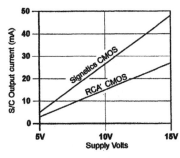

Figure 4.23 Typical '4000B'-series short-circuit output currents (at 25°C)

'4000B'-series CMOS output stages over the 5V to 15V operating voltage range; in practice, the maximum DC values of these output loads must be limited to 10mA of current or 100mW of power dissipation, whichever is the lower of these values.

Figure 4.24 shows the typical short-circuit output currents of 'standard' and 'bus driver' versions of '74HC'-series CMOS output stages over the 2V to 6V operating voltage range; in practice, the maximum DC values of these currents must be limited to 25mA in 'standard' HC types, and 35mA in 'bus driver' HC types.

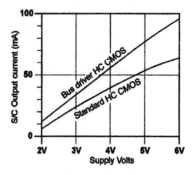

Figure 4.24   Typical '74HC'-series short-circuit output currents (at 25°C)

The only time this 'current limiting' matter is likely to present any real problem is when using CMOS to drive some type of LED load (including those at the inputs of optocouplers, etc.). *Figures 4.25* and *4.26* show basic ways of driving an LED via non-inverting or inverting CMOS elements. Note in these circuits that R1 sets the LED's ON current, and has a value of $[(V+ - V_s)/I] -$ Rx, where V+ is the supply voltage, V_s is the LED's saturation voltage (typically 2.0 to 2.5 volts), I is the LED's ON current (in amps), and Rx is the CMOS elements saturation resistance (and varies widely with voltage, current, and with individual ICs). Typically, however, Vs equals 2.2V, and Rx has an approximate value of 100R in a standard 74HC output or 70R in a bus driver output, or 500R in a standard 4000B output. Thus, to set the LED current at 10mA, R1 needs a value

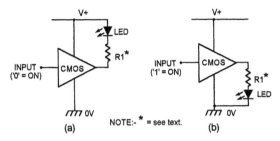

Figure 4.25 LED-driving output interface, using non-inverting CMOS elements

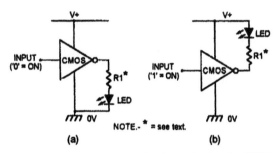

Figure 4.26 LED-driving output interface, using inverting CMOS elements

of about 180R in a 5V standard 74HC circuit, 220R in a 5V bus driver 74HC circuit, 270R in a 10V 4000B circuit, or 820R in a 15V 4000B circuit.

Note that CMOS outputs can be used to drive any of the basic 'TTL' output interface circuits shown in *Figures 3.12* to *3.17* (in Chapter 3) by simply wiring a current-limiting resistor in series with the CMOS output, to limit its output current to a safe value.

Logic family interfacing
It is generally bad practice to mix different logic families in any system, but on those occasions where it does occur the mix is usually made between TTL and CMOS devices; *Figures 3.18* to *3.23* of Chapter 3 showed six basic ways of interfacing TTL and CMOS ICs. Note that 74HCT and 74ACT types of CMOS IC are designed to be directly driven from TTL outputs, without the need for special interfacing methods. Also

note that standard '4000B'-series and '74CXX'-series CMOS elements have very low fan-outs and can only drive a single Standard TTL or LS TTL element, but '74HCXX'-series (and '74ACXX'-series) CMOS elements have excellent fan-outs and can directly drive up to 2 Standard TTL inputs, or 10 LS TTL inputs, or 20 ALS TTL inputs.

Most TTL ICs with open-collector (o.c.) outputs have output-voltage ratings of at least 15V (but the main IC has a normal 5V rating), and can be interfaced to the input of a CMOS logic IC by using the connections shown in *Figure 4.27*. Here, R1 acts like a pull-up resistor, and the CMOS IC can either share the 5V supply of the TTL IC, or can use its own 5V to 15V positive supply rail. Similarly, a CMOS IC with an open-drain (o.d.) output can be interfaced to a normal TTL input by using the connections shown in *Figure 4.28*, but in this case the two ICs must share a common 5V supply rail.

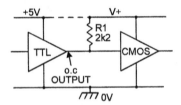

Figure 4.27 TTL (o.c. output) to CMOS interface, using common or independent +ve rails

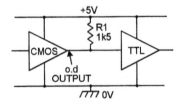

Figure 4.28 CMOS (o.d. output) to TTL interface

CMOS Supply Pin Notations

Most digital ICs have only two supply pins, one of which connects to a circuit's positive supply rail, and the other to the zero-volts rail. In TTL ICs, these pins are conventionally notated V_{CC} and GND respectively, with the 'V_{CC}' notation implying that the positive rail usually connects to the collector sides of the IC's internal transistors. When '4000'-series CMOS ICs were first introduced, its supply pins were renamed V_{DD} and V_{SS} respectively, implying that the positive rail usually connects to the drain side of the IC's internal IGFETs, and the zero-volts rail to the source sides; these notations are in fact totally ambiguous, as can be seen from the basic CMOS circuit of *Figure 4.1*, which shows that both rails in fact connect to IGFET sources. These notations are, however, still often used in CMOS manufacturers' data books.

When CMOS was first used as a 'C' sub-family in the '74'-series range of ICs, its supply pins were renamed V_{CC} and GND, to comply with normal TTL conventions, and this system has subsequently been used on all other CMOS sub-families used in the '74'-series of ICs. In recent times this same system has started to be used on the '4000'-series of CMOS ICs as well, and it is used throughout most of this book, with the major exceptions of Chapter 5, which is devoted to one of the oldest but most versatile of all CMOS ICs, the 4007UB, and Chapter 7, which describes CMOS bilateral switch ICs. In Chapter 5, the CMOS IC's positive supply pin is notated V_{DD}, and the zero-volts pin is notated GND. In Chapter 7, supply pin notations of V_{DD} and V_{SS} are used.

5 Using the 4007UB CMOS IC

If you are unfamiliar with modern CMOS, the best way to learn about it is by experimenting with the inexpensive 4007UB IC. The 4007UB houses little more than two pairs of complementary MOSFETs and one simple CMOS inverter stage; all of these elements are, however, independently accessible, and can be configured in a variety of ways. The 4007UB is thus a very versatile IC, and is ideal for demonstrating CMOS principles to students, technicians, and engineers. It can readily be configured to act as a multiple digital inverter, a NAND or NOR gate, a transmission gate, or a uniquely versatile 'micropower' linear amplifier or oscillator, etc. This chapter presents a selection of practical circuits of these types; it starts off, however, by outlining 4007UB basics.

4007UB Basics

Figure 5.1(a) shows the functional diagram and pin numbering of the 14-pin 4007UB, which houses two complementary pairs of independently accessible MOSFETs, plus a third complementary pair that is connected in the form of a basic CMOS inverter stage. Each of the three independent input terminals of the IC is internally connected to the standard CMOS protection network shown in *Figure 5.1(b)*. All MOSFETs in the 4007UB are enhancement-mode devices; Q1, Q3 and Q5 are p-channel types, and Q2, Q4 and Q6 are n-channel types. *Figure 5.1(c)* shows the terminal notations of the two MOSFET types; note that the 'B' terminal represents the bulk substrate. All modern '4000B'-series and fast '74'-series CMOS ICs are designed around the basic elements shown in *Figure 5.1*, and it is thus useful to get a good basic understanding of both the digital and the linear characteristics of these elements, starting off with those of the basic MOSFETs.

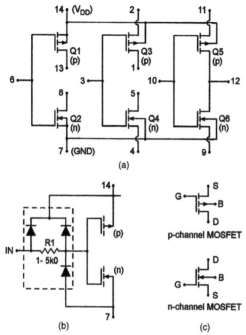

Figure 5.1 *(a)* Functional diagram of the 4007UB dual CMOS pair plus inverter *(b)* Internal input-protection network (within dotted lines) on each input of the 4007UB *(c)* MOSFET terminal notations; G = Gate, D = Drain, S = Source, B = Bulk substrate

Digital operation

The input (gate) terminal of a MOSFET presents a near-infinite impedance to DC voltages, and the magnitude of an external voltage applied to the gate controls the magnitude of the MOSFET's source-to-drain current flow. The basic characteristics of the enhancement-mode n-channel MOSFET are such that the source-to-drain path is open-circuit when the gate is at the same potential as the source, but becomes a near short-circuit (a low-value resistance) when the gate is heavily biased *positive* to the source. Thus, the n-channel MOSFET can be used as a digital inverter by wiring it as shown in *Figure 5.2*; with a logic-0 (zero-volts) input the MOSFET is cut off and the output is at logic-1 (the positive rail voltage), but with a logic-1 input the MOSFET is driven on and the output is at logic-0.

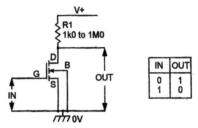

Figure 5.2   Digital inverter made from n-channel MOSFET

The basic characteristics of the enhancement-mode p-channel MOSFET are such that the source-to-drain path is open when the gate is at the same potential as the source, but becomes a near-short when the gate is heavily biased *negative* to the source. The p-channel MOSFET can thus be used as a digital inverter by wiring it as shown in *Figure 5.3*.

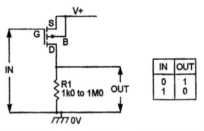

Figure 5.3   Digital inverter made from p-channel MOSFET

Note in the *Figure 5.2* and *5.3* inverter circuits that the ON currents of the MOSFETs are determined by the R1 value, and these circuits thus draw a significant quiescent current when their MOSFETs are driven ON. This snag can be overcome by connecting the complementary pair of MOSFETs in the classic CMOS inverter configuration shown in *Figure 5.4(a)*.

In *Figure 5.4(a)*, with a logic-0 input applied, Q1 is driven fully on and the output is thus firmly tied to the logic-1 (positive rail) state, but Q2 is cut off and the inverter thus passes zero quiescent current via this MOSFET. With a logic-1 input applied, Q2 is driven on and the output is firmly tied to the logic-0 (zero-volt) state, but Q1 is cut off

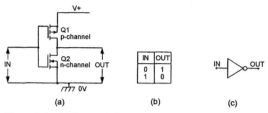

Figure 5.4 *(a)* Circuit, *(b)* Truth Table and *(c)* standard symbol of the CMOS digital inverter

and the circuit again passes zero quiescent current. This 'zero quiescent current' characteristic of the complementary MOSFET inverter is one of the most important features of the CMOS digital inverter, and the *Figure 5.4(a)* circuit forms the basis of the entire CMOS family of digital ICs. Q5 and Q6 of the 4007UB are fixed-wired in this CMOS inverter configuration.

Linear operation

To understand fully the operation and vaguaries of CMOS circuitry, it is necessary to understand the linear characteristics of basic MOSFETs. *Figure 5.5* shows the typical gate-voltage/drain-current graph of an n-channel enhancement mode MOSFET. Note that negligible drain current flows until the gate voltage rises to a 'threshold' value of about 1.5 to 2.5 volts, but that the drain current then increases almost linearly with further increases in gate voltage.

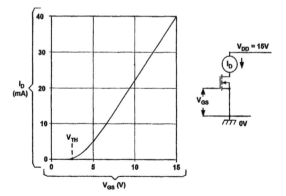

Figure 5.5 Typical gate-voltage/drain-current characteristics of an n-channel MOSFET

Figure 5.6 shows how to connect an n-channel 4007UB MOSFET as a linear inverting amplifier. R1 serves as the drain load of Q2 and R2–Rx bias the gate so that the device operates in the linear mode. The Rx value must be selected to give the desired quiescent drain voltage, but is normally in the range 18k to 100k. The amplifier can be made to give a very high input impedance by wiring a 10M isolating resistor between the R2–Rx junction and the gate of Q2, as shown in *Figure 5.6(b)*.

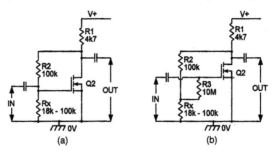

Figure 5.6 Methods of biasing an n-channel 4007UB MOSFET as a linear inverting amplifier

Figure 5.7 shows the typical I_D/V_{DS} characteristics of an n-channel MOSFET at various fixed values of gate-to-source voltage. Imagine here that, for each set of curves, V_{GS} is fixed at the V_{DD} voltage, but that the V_{DS} output voltage can be varied by altering the

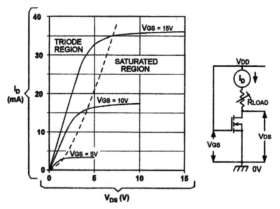

Figure 5.7 Typical I_D to V_{DS} characteristics of the n-channel MOSFET at various fixed values of V_{GS}

value of drain load R_L. The graph can be divided into two characteristic regions, as indicated by the dotted line, these being the *triode* region and the *saturated* region.

When the MOSFET is in the saturated region (with V_{DS} at some value in the nominal range 50 to 100% of V_{GS}) the drain acts like a constant current source, with its current value controlled by V_{GS}: a low V_{GS} value gives a low constant-current value, and a high V_{GS} value gives a high constant-current value. These saturated constant-current characteristics provide CMOS with an output 'short-circuit proof' feature and also determine its operating speed limits at different supply voltage values.

When the MOSFET is in the triode region (with V_{DS} at some value in the nominal range 1 to 50% of V_{GS}) the drain acts like a voltage-controlled resistance, with the resistance value increasing approximately as the square of the V_{GS} value.

The p-channel MOSFET has an I_D/V_{DS} characteristic graph that is complementary to that of *Figure 5.7*. Consequently, the action of the basic CMOS inverter of *Figure 5.4* (which uses a complementary pair of MOSFETs) is such that its current-drive capability into an external load, and also its operating speed limits, increase in proportion to the supply rail voltage.

Figure 5.8 shows the typical voltage-transfer characteristics of the 4007UB's standard CMOS inverter

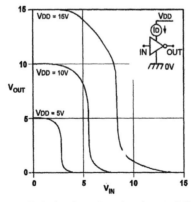

Figure 5.8 Typical voltage transfer characteristics of the 4007UB simple CMOS inverter

at different supply voltage values. Note (on the 15V V_{DD} line, for example) that the output voltage changes by only a small amount when the input voltage is shifted around the V_{DD} and 0V levels, but that when V_{in} is biased at roughly half-supply volts a small change of input voltage causes a large change of output voltage: typically, the inverter gives a voltage gain of about 30dB when used with a 15 volt supply, or 40dB at 5 volts. *Figure 5.9* shows how to connect the CMOS inverter as a linear amplifier; the circuit has a typical bandwidth of 710kHz at 5 volts supply, or 2.5MHz at 15 volts.

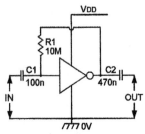

Figure 5.9   Method of biasing the simple CMOS inverter for linear operation

Wiring three simple CMOS inverter stages in series as in *Figure 5.10(a)* gives the direct equivalent of a modern '4000B'-series 'buffered' CMOS inverter stage, which has the overall voltage transfer graph of *Figure 5.10(b)*. The 'B'-series inverter typically gives about

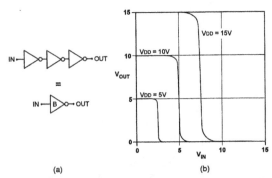

(a)   (b)

Figure 5.10   Wiring three simple CMOS inverters in series *(a)* gives the equivalent of a 'B'-series 'buffered' CMOS inverter, which has the transfer characteristics shown in *(b)*

70dB of linear voltage gain, but tends to be grossly unstable when used in the linear mode.

Finally, *Figure 5.11* shows the drain-current transfer characteristics of the simple CMOS inverter. Note that the drain current is zero when the input is at either zero or full supply volts, but rises to a maximum value (typically 0.5mA at 5V supply, or 10.5mA at 15V supply) when the input is at roughly half-supply volts, under which condition both MOSFETs of the inverter are biased on. In the 4007UB, these ON currents can be reduced by wiring extra resistance in series with the source of each MOSFET of the CMOS inverter; this technique is used in the 'micropower' circuits shown later in this chapter.

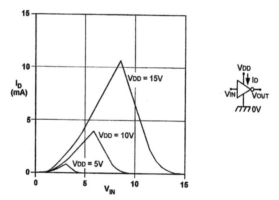

Figure 5.11   Drain-current transfer characteristics of the simple CMOS inverter

Using the 4007UB

The 'usage' rules of the 4007UB are quite simple. In any specific application, all unused elements of the device must be disabled; complementary pairs of MOSFETs can be disabled by connecting them as standard CMOS inverters and tying their inputs to ground, as shown in *Figure 5.12*; individual MOSFETs can be disabled by tying their source to their substrate (B) and leaving the drain open-circuit.

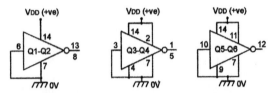

Figure 5.12 Individual 4007UB complementary MOSFET pairs can be disabled by connecting them as CMOS inverters and grounding their inputs

In use, the input terminals must not be allowed to rise above V_{DD} (the supply voltage) or below V_{SS} (zero volts). To use an n-channel MOSFET, the source must be tied to V_{SS}, either directly or via a current-limiting resistor. To use a p-channel MOSFET, the source must be tied to V_{DD}, either directly or via a current-limiting resistor.

Practical Circuits: Digital

The 4007UB elements can be configured to act as any of a variety of standard digital circuits. *Figure 5.13* shows how to wire the IC as a triple inverter, using all three sets of complementary MOSFET pairs. *Figure 5.14* shows the connections for making an inverter plus non-inverting buffer; here, the Q1–Q2 and Q3–Q4 inverter stages are simply wired directly in series, to give an overall non-inverting action.

The maximum source and sink output currents of a simple CMOS inverter stage self-limit at about 10–20mA as one or other of the output MOSFETs turns fully on. Higher sink currents can be obtained by simply

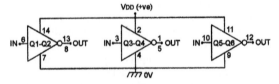

Figure 5.13   4007UB triple inverter

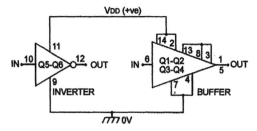

Figure 5.14 4007UB inverter plus non-inverting buffer

wiring n-channel MOSFETs in parallel in the output stage. *Figure 5.15* shows the 4007UB wired so that it acts as a high-sink-current inverter that will absorb triple the current of a normal inverter. Similarly, *Figure 5.16* shows how to wire the 4007UB to act as a high-source-current inverter, and *Figure 5.17* shows the connections for making a single inverter that will sink or source three times more current than a standard inverter stage.

The 4007UB is an ideal device for demonstrating basic CMOS logic gate principles. *Figure 5.18* shows it used

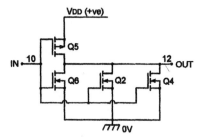

Figure 5.15 4007UB high sink-current inverter

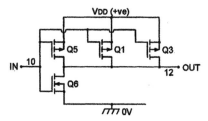

Figure 5.16 4007UB high source-current inverter

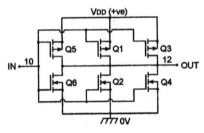

Figure 5.17 4007UB high-power inverter, with triple the sink- and source-current capability of a standard inverter

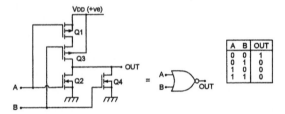

Figure 5.18 4007UB 2-input NOR gate

for making a 2-input NOR gate; note that the two n-channel MOSFETs are wired in parallel so that either can pull the output to ground from a logic-1 input, and the two p-channel MOSFETs are wired in series so that both must turn on to pull the output high from a logic-0 input. The Truth Table shows the logic of the circuit. A 3-input NOR gate can be made by simply wiring three p-channel MOSFETs in series and three n-channel MOSFETs in parallel, as shown in *Figure 5.19*.

Figure 5.20 shows the 4007UB used as a 2-input NAND gate, with the two p-channel MOSFETs wired in parallel and the two n-channel MOSFETs wired in series. A 3-input NAND gate can be made by similarly wiring three p-channel MOSFETs in parallel and three n-channel MOSFETs in series.

Figure 5.21 shows the basic way of using the 4007UB to make another important CMOS element, the transmission gate or bilateral switch, which acts like a near-perfect switch that can conduct signals in either direction and can be turned on (closed) by applying a logic-1 to its control

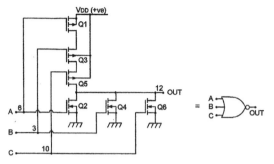

Figure 5.19 4007UB 3-input NOR gate

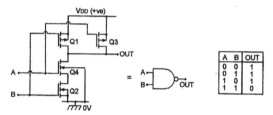

A	B	OUT
0	0	1
0	1	1
1	0	1
1	1	0

Figure 5.20 4007UB 2-input NAND gate

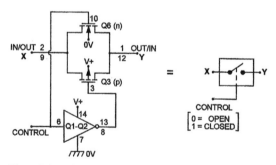

Figure 5.21 4007UB transmission gate or bilateral switch

terminal or turned off (open) via a logic-0 control signal. In *Figure 5.21*, an n-channel and a p-channel MOSFET are wired in parallel (source-to-source and drain-to-drain), but their gate signals are applied in anti-phase via the Q1–Q2 inverter. To turn the Q3–Q6 transmission gate on (closed), Q6 gate is taken to logic-1 and Q3 gate to logic-0 via the inverter; to turn the switch off, the gate polarities are simply reversed. The 4007UB transmission

gate has a near-infinite OFF resistance and an ON
resistance of about 600R. It can handle all signals
between zero volts and the positive supply rail value.
Note that, since the gate is bilateral, either of its main
terminals can function as an input or output.

Finally, *Figure 5.22* shows how the 4007UB can be wired
as a dual transmission gate that functions like a single-pole
double-throw (SPDT) switch. In this case the circuit uses
two transmission elements, but their control voltages are
applied in anti-phase, so that one switch opens when the
other closes, and vice versa. The 'X' sides of the two gates
are shorted together to give the desired SPDT action.

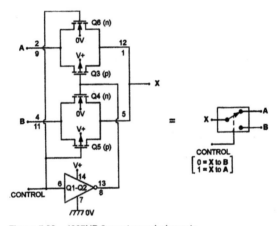

Figure 5.22   4007UB 2-way transmission gate

Practical Circuits: Linear

Figures 5.6 to *5.9* have already shown that the basic
4007UB MOSFETs and the CMOS inverter can be used
as linear amplifiers. *Figure 5.23* shows the typical
voltage gain and frequency characteristics of the linear
CMOS inverter when operated from three alternative
supply rail values (this graph assumes that the amplifier
output is feeding into the high impedance of a
10M/15pF oscilloscope probe). The output impedance

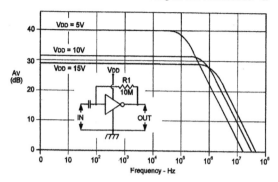

Figure 5.23 Typical Av and frequency characteristics of the linear-mode basic CMOS amplifier

of the open-loop amplifier typically varies from 3k0 at 15 volts supply, to 5k0 at 10 volts, or 22k at 5 volts, and it is the product of the output impedance and output load capacitance that determines the bandwidth of the circuit; increasing the output impedance or load capacitance reduces the bandwidth.

As you would expect from the voltage transfer graph of *Figure 5.8*, the distortion characteristics of the CMOS linear amplifier are not very good. Linearity is fairly good for small-amplitude signals (output amplitudes up to 3 volts pk-to-pk with a 15V supply), but the distortion then increases progressively as the output approaches the upper and lower supply limits. Unlike a bipolar transistor circuit, the CMOS amplifier does not 'clip' excessive sine wave signals, but progressively rounds off their peaks.

Figure 5.24 shows the typical drain-current/supply-voltage characteristics of the basic CMOS linear

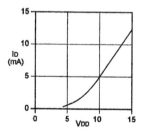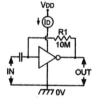

Figure 5.24 Typical I_D/V_{DD} characteristics of the linear-mode CMOS amplifier

amplifier. Note that the supply current typically varies from 0.5mA at 5 volts to 12.5mA at 15 volts.

'Micropower' Circuits

In many applications, the quiescent supply current of the 4007UB CMOS linear amplifier can be usefully reduced, at the expense of reduced amplifier bandwidth, by wiring external resistors in series with the source terminals of the two MOSFETs of the CMOS stage, as shown in the 'micropower' circuit of *Figure 5.25*. This diagram also shows the effect that different resistor values have on the drain current, voltage gain and bandwidth of the amplifier when it is operated from a 15 volt supply and has its output feeding to a 10M/15pF oscilloscope probe.

It is important to appreciate in the *Figure 5.25* circuit that these additional resistors add to the output impedance of the amplifier (the output impedance roughly equals the R1–Av product), and this impedance and the external load resistance/capacitance has a great effect on the overall gain and bandwidth of the circuit. When using 10k values for R1, for example, if the load capacitance is increased to 50pF the bandwidth falls to about 4kHz, but if the capacitance is reduced to 5pF the bandwidth rises to 45kHz. Similarly, if the resistive load

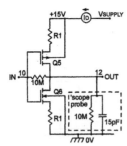

R1	I$_D$	Av (V$_{OUT}$/V$_{IN}$)	Upper 3dB Bandwidth
0	12.5mA	20	2.7MHz
100R	8.2mA	20	1.5MHz
560R	3.9mA	25	300kHz
1k0	2.5mA	30	150kHz
5k6	600µA	40	25kHz
10k	370µA	40	15kHz
100k	40µA	30	2kHz
1M0	4µA	10	1kHz

Figure 5.25 'Micropower' 4007UB CMOS linear amplifier, showing method of reducing I$_D$, with measured performance details

is reduced from 10M to 10k, the voltage gain falls to unity. Thus, for significant gain, the load resistance must be large relative to the output impedance of the amplifier.

The basic (unbiased) CMOS inverter stage has an input capacitance of about 5pF and an input resistance of near infinity. Thus, if the output of the *Figure 5.25* circuit is fed directly to such a load, it will show a voltage gain of about ×30 and a bandwidth of 3kHz when R1 has a value of 1M0; it will even give useful gain and bandwidth when R1 has a value of 10M, but will consume a quiescent current of only 0.4μA!

The CMOS linear amplifier can be used, in either its standard or micropower forms, to make a variety of fixed-gain amplifiers, mixers, integrators, active filters and oscillators, etc. Three typical basic applications are shown in *Figure 5.26*.

One attractive 4007UB linear application is as a crystal oscillator, as shown in *Figure 5.27(a)*. Here, the CMOS amplifier is linearly biased via R1 and provides 180° phase shift, and the Rx–C1–XTAL–C2 pi-type crystal

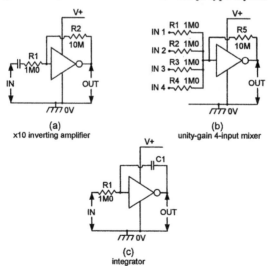

Figure 5.26 The CMOS amplifier can be used in a variety of linear inverting amplifier applications. Three typical examples are shown here

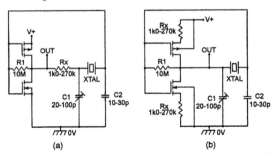

Figure 5.27 Crystal oscillator using *(a)* standard and *(b)* micropower 4007UB CMOS linear inverter

network gives an additional 180° of phase shift at the crystal resonant frequency, thereby causing the circuit to oscillate. If this circuit is needed to provide a frequency accuracy within only 0.1% or so, Rx can be replaced by a short and C1–C2 can be omitted; for ultra-high accuracy, the correct values of Rx–C1–C2 must be individually determined (*Figure 5.27* shows the typical range of values). In micropower applications, Rx can be incorporated in the CMOS amplifier, as shown in *Figure 5.27(b)*. If desired, the output of the crystal oscillator can be fed directly to the input of an additional CMOS inverter stage, for improved waveform shape/amplitude.

Practical Circuits: Astables

One of the most useful applications of the 4007UB is as a ring-of-three astable multivibrator; *Figure 5.28* shows the basic configuration of the circuit. Waveform timing is controlled by the values of R1 and C1, and the output waveform (A) is approximately symmetrical. Note that for most of the waveform period the front-end (waveform (B)) part of the circuit operates in the linear mode, so the circuit consumes a significant running current.

In practice, the running current of the *Figure 5.28* astable circuit is higher than that of an identically

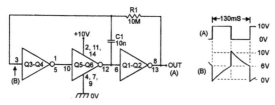

Figure 5.28   This 4007UB ring-of-three astable consumes 280µA at 6V, 1.6mA at 10V

configured B-series 'buffered' CMOS IC such as the 4001B, the comparative figures being 280µA at 6V or 1.6mA at 10V for the 4007UB, against 12µA at 6V or 75µA at 10V for the 4001B. The 4007UB circuit, however, has far lower propagation delays than the 4001B and typically has a maximum astable operating speed that is three times higher than that of the 4001B.

The running current of the 4007UB astable can be greatly reduced by operating its first two stages in the 'micropower' mode, as shown in *Figure 5.29*. This technique is of special value in low-frequency operation, and the *Figure 5.29* circuit in fact consumes a mere 1.5µA at 6V or 8µA at 10V, these figures being far lower than those obtainable from any other IC in the CMOS range. The frequency stability of the *Figure 5.29* circuit is not, however, very good, the period varying from 200ms at 6V to 80mS at 10V.

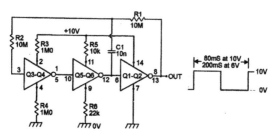

Figure 5.29   This micropower ring-of-three symmetrical 4007UB astable consumes 1.5µA at 6V, 8µA at 10V

Figure 5.30 shows the 4007UB wired as an asymmetrical ring-of-three astable in which the circuit's 'input' is applied to n-channel MOSFET Q2; this circuit

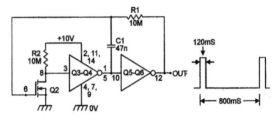

Figure 5.30   This 4007UB asymmetrical ring-of-three astable consumes 2μA at 6V, 5μA at 10V

consumes a mere 2μA at 6V or 5μA at 10V. *Figure 5.31* shows how the symmetry of the circuit can be varied by shunting R1 with the D1–R3 network, so that the charge and discharge times of C1 are independently controlled; with the component values shown, this circuit produces a 300μS pulse once every 900ms and consumes 2μA at 6V or 4.5μA at 10V.

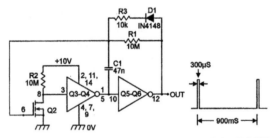

Figure 5.31   This dual-time-constant version of the 4007UB astable generates a very narrow output pulse

Finally, *Figure 5.32* shows how the *Figure 5.31* circuit's current consumption can be further reduced by

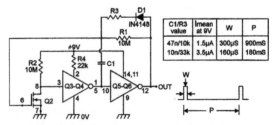

C1/R3 value	Imean at 9V	W	P
47n/10k	1.5μA	300μS	900mS
10n/33k	3.5μA	160μS	180mS

Figure 5.32   This micropower version of the 4007UB dual-time-constant astable consumes absolutely minimal currents

operating the Q3–Q4 CMOS inverter in the micropower mode. The table gives details of circuit performance with alternative C1 and R3 values. This circuit can give years of continuous operation from a small 9V battery.

6 Buffer, Gate, and Logic IC Circuits

Earlier chapters of this book explained TTL and CMOS principles and basic usage rules, and looked at the '74'- and '4000'-series of ICs and at various sub-members of these families. The present chapter carries on the 'digital ICs' theme by showing practical ways of using TTL and CMOS buffers, inverters, and logic gate ICs.

Logic Gate Symbology

The eight best known types of logic 'gate' are the buffer and the NOT, OR, NOR, AND, NAND, EX-OR and EX-NOR types. Many different symbols can be used to represent each of these eight basic logic gate elements. *Figure 6.1* shows four different families of symbols that

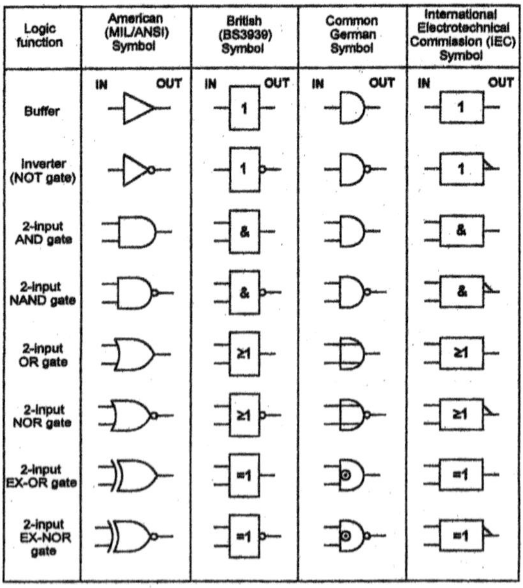

Logic function	American (MIL/ANSI) Symbol	British (BS3939) Symbol	Common German Symbol	International Electrotechnical Commission (IEC) Symbol
Buffer		1		1
Inverter (NOT gate)		1		1
2-input AND gate		&		&
2-input NAND gate		&		&
2-input OR gate		≥1		≥1
2-input NOR gate		≥1		≥1
2-input EX-OR gate		=1		=1
2-input EX-NOR gate		=1		=1

Figure 6.1 A selection of widely used logic symbols

are widely used in various parts of the world today; of these, the American MIL/ANSI symbols are by far the most popular, are instantly recognizable, are used by most of the world's practical digital engineers, and are used throughout this book. Two useful variations of these American symbols are also widely used and are shown added to a standard inverter symbol in *Figure 6.2*; the left-hand symbol is internationally recognized and indicates that the logic element has a Schmitt-trigger input action; the right-hand symbol – which is widely used but is not universally recognized – indicates that the logic element has an open-drain (o.d.) or open-collector (o.c.) output stage.

Schmitt inverter   **Inverter with o.c. (open collector) or o.d. (open drain) output**

Figure 6.2 Useful variations of the MIL/ANSI inverter symbol

Logic Gate Functions

The functional action of any logic gate can be described either in words or in a tabular or symbolic way. The following list describes the functions of all eight basic types of gate in words.

Buffers. A buffer is a non-inverting amplifier that has an output drive capacity that is far greater than its input drive requirement, i.e. it has a high fan-out and gives a logic-1 output for a logic-1 input, etc.

Inverters. An inverter (also known as a NOT gate) is a high fan-out amplifier that gives a logic-1 output for a logic-0 input, and vice versa.

AND gates. An AND gate has an output that is normally at logic-0 and only goes to logic-1 when *all* inputs are at logic-1, i.e. when inputs A *and* B *and* C, etc. are high.

NAND gates. A NAND gate is an AND gate with a negated (inverted) output; the output is normally at logic-1 and only goes to logic-0 when *all* inputs are at logic-1.

OR gates. An OR gate has an output that goes to logic-1 if *any* input is at logic-1, i.e. if inputs A *or* B *or* C, etc., are high. The output goes to logic-0 only if *all* inputs are at logic-0.

NOR gates. A NOR gate is an OR gate with a negated output; it has an output that goes to logic-0 if any input is at logic-1, and goes to logic-1 only when *all* inputs are at logic-0.

EX-OR gates. An exclusive-OR (EX-OR) gate has two inputs, and its output goes to logic-1 only if a *single* input (A *or* B) is at logic-1; the output goes to logic-0 if *both* inputs are in the same logic state.

EX-NOR gates. An exclusive-NOR (EX-NOR) gate is an EX-OR gate with a negated output, which goes to logic-1 if *both* inputs are in the same logic state, and goes to logic-0 only if a single input is at logic-1.

Figure 6.3 shows how the functions of the eight gates can also be presented in tabular form via Truth Tables (which show the logic state of the output at all possible input logic state combinations), or symbolically via Boolean algebraic terms (an explanation of Boolean algebra basics is given in the author's *Modern TTL Circuits Manual*). Note that, by convention, all logic gate inputs are notated alphabetically as 'A', 'B', 'C', etc., and the output terminal is notated as 'Y' (but in counters and flip-flops, etc., the main output is usually notated as 'Q'); the actual logic states may be represented by '0' and '1', as shown, or by 'L' (= Low logic level) and 'H' (= High logic level). Also note in the Boolean expressions that a negated output is indicated by a negation bar drawn above the basic output symbol; the negated state is called a 'not' state; thus, a negated 'Y' output is called a 'not-Y' output.

Positive Versus Negative Logic

All modern digital logic circuitry assumes the use of the 'positive logic' convention, in which a logic-1 state is High and a logic-0 state is Low. In the early days of

Logic function	Logic symbol	Truth table	Boolean expression
Buffer	 A ▷ Y	A Y 0 0 1 1	$Y = A$
Inverter (NOT gate)	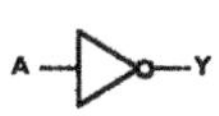 A ▷o Y	A Y 0 1 1 0	$Y = \overline{A}$
2-input AND gate	 A, B	A B Y 0 0 0 0 1 0 1 0 0 1 1 1	$Y = A \cdot B$
2-input NAND gate	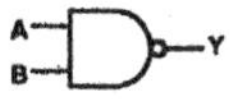 A, B	A B Y 0 0 1 0 1 1 1 0 1 1 1 0	$Y = \overline{A \cdot B}$
2-input OR gate	 A, B	A B Y 0 0 0 0 1 1 1 0 1 1 1 1	$Y = A + B$
2-input NOR gate	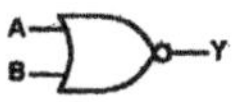 A, B	A B Y 0 0 1 0 1 0 1 0 0 1 1 0	$Y = \overline{A + B}$
2-input EX-OR gate	 A, B	A B Y 0 0 0 0 1 1 1 0 1 1 1 0	$Y = A \oplus B$
2-input EX-NOR gate	 A, B	A B Y 0 0 1 0 1 0 1 0 0 1 1 1	$Y = \overline{A \oplus B}$

Figure 6.3 Symbols, Truth Tables, and Boolean expressions for eight basic types of logic gate

electronic digital circuitry an alternative 'negative logic' convention – in which a logic-1 state is Low and a logic-0 state is High – was also in common use, and it is sometimes still useful to be able to think in negative-logic terms, particularly when designing gates in which a Low-state output is of special interest. With this point in mind, *Figure 6.4* presents a basic set of 2-input Positive and Negative logic equivalents. Thus, it can be

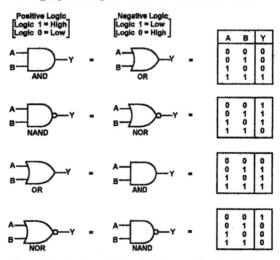

Figure 6.4   Basic sets of 2-input Positive and Negative logic equivalents

seen that a Negative logic AND gate action – in which the output is Low only when both inputs are Low – is directly available from a Positive logic OR gate, and so on.

Practical Buffer IC Circuits

Digital buffer ICs have two main purposes, and these are to act either as simple non-inverting current-boosting interfaces between one part of a circuit and another, or to act as 3-state switching units that can be used to connect a circuit's outputs to a load only when required. If you ever need only a few simple buffers, one cheap way to get them is to make them from spare AND or OR elements, as shown in *Figure 6.5*, or from pairs of normal or Schmitt inverters, as shown in *Figure 6.6*.

Figure 6.5   Any AND or OR gate can be used as a non-inverting buffer element

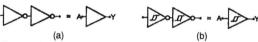

(a)   (b)

Figure 6.6   Any two elements from an inverter IC can be used to make a non-inverting buffer element

Figure 6.7 lists basic details of nine popular non-inverting buffer ICs. When using these ICs, note that all unused buffers must be disabled by tying their inputs to one of the IC's supply lines. In CMOS devices, the unused inputs can be tied directly to either supply line, but in TTL devices it is best (for lowest quiescent current consumption) to tie all unused inputs high via a common 10k resistor; if the unused buffer is a 3-state type, it should (if it has independent control) be set into its 'normal' mode via its control input.

Device	Type	Description
74LS125	LS TTL	Quad 3-state buffer
4050B	CMOS	Hex buffer
4503B	CMOS	Hex (Dual + Quad) 3-state buffer
7407	TTL	Hex buffer with 30V o.c. outputs
74LS365	LS TTL	Hex 3-state buffer
74HC4050	CMOS	Hex buffer
74HC241	CMOS	Octal (Dual Quad) 3-state Schmitt buffer
74HC244	CMOS	Octal (Dual Quad) 3-state Schmitt buffer
74LS244	LS TTL	Octal (Dual Quad) 3-state Schmitt buffer

Figure 6.7   Nine popular non-inverting buffer ICs

Dealing now with the individual buffer ICs listed in *Figure 6.7*, *Figure 6.8* shows the functional diagram and Truth Table, etc., of the 74LS125 TTL IC, which houses four independently controlled 3-state buffers and is so modestly priced that it is still worth using even if you do not need the '3-state' facility. Note from the Truth Table that each of the four elements acts as a normal buffer when its control terminal (C) is in the logic-0 state, and that the element's quiescent current (I_s) is least when C

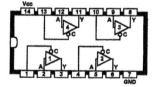

C	A	Y	Is (per gate)
0	0	0	2.25mA
0	1	1	1.12mA
1	0	HI-Z	2.75mA
1	1	HI-Z	2.62mA

Note: Normal fan-out = 30

Figure 6.8   Functional diagram and Truth Table, etc., of the 74LS125 Quad 3-state buffer/bus-driver IC

is at logic-0 and the buffer's input (A) is at logic-1. Thus, any unwanted elements should be disabled by tying their C terminals Low and their A terminals High, using one of the methods shown in *Figure 6.9*, and any element can be used as a normal buffer by grounding its C terminal as in *Figure 6.10*, or as a 3-state buffer that drives a common bus line by using it as shown in *Figure 6.11*.

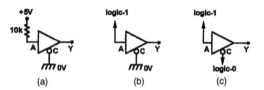

(a)　　　　　(b)　　　　　(c)

Figure 6.9   All unwanted 74LS125 elements must be connected in one of these ways

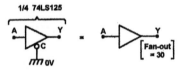

Figure 6.10   Methods of using a 74LS125 element as a normal buffer

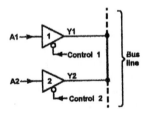

Figure 6.11   Methods of using a 74LS125 element as a 3-state line-driving buffer

If you need up to six simple CMOS buffers, one of the cheapest ways to get them is via a 4050B or 74HC4050 Hex buffer IC. *Figure 6.12* shows the IC's functional diagram; each buffer can source up to 10mA or sink up to 40mA of output current when the IC is powered from a 15V supply.

If you need up to six 3-state CMOS buffers, one option is to use a 4503B Hex buffer IC. *Figure 6.13* shows the

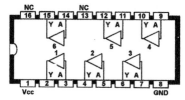

Figure 6.12   Functional diagram of the 4050B or 74HC4050 Hex buffer IC

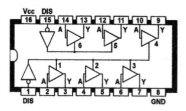

IN (A)	DISable input	OUT (Y)
0	0	0
1	0	1
X	1	Hi-Z

X = Don't care

Figure 6.13   Functional diagram and Truth Table of the 4503B Hex buffer IC

functional diagram and Truth Table of this versatile IC, in which pin-1 acts as a DISABLE input that controls four of the six buffers, and pin-15 acts as a DISABLE input that controls the other two buffers. Note that each buffer element acts as a normal buffer when its DISABLE pin is at logic-0 (low), and goes into the high-impedance output state when its DISABLE pin is at logic-1 (high). Thus, this IC can be used as a simple Hex buffer by wiring it as shown in *Figure 6.14* (with

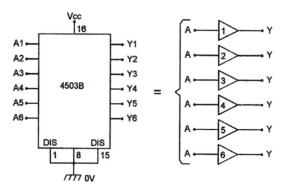

Figure 6.14   Method of connecting the 4503B for use as six normal buffers

pins 1 and 15 grounded), or as a Hex 3-state buffer that
is controlled via a single input by wiring it as shown in
Figure 6.15 (with pins 1 and 15 shorted together and
used as a DISABLE input)

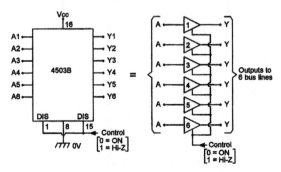

Figure 6.15   Method of connecting the 4503B for use as a 3-
state Hex buffer controlled via a single input

Figure 6.16 shows the functional diagram of the 7407,
which is a Standard-TTL Hex buffer in which each
buffer has an open-collector (o.c.) output that can sink
up to 40mA and can be connected to a supply of up to
30V via an external current-limiting pull-up resistor (but
the actual IC must use a 5V supply); *Figure 6.17* shows

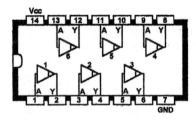

Figure 6.16   Functional diagram of the 7407 Hex buffer with 30V
o.c. outputs

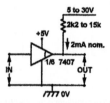

Figure 6.17   5V to high-voltage buffer/interface

how one of these buffers can be used as a 5V to high-
voltage (up to 30V) non-inverting interface. *Figure 6.18*
shows how three o.c. buffers can be made to act as a
wired-AND gate by wiring all three outputs to the same
pull-up resistor; the circuit action is such that the output
is pulled low when any input is low, and only goes high
when all three inputs are high, thus giving an AND
action.

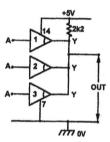

Figure 6.18   Three 7407 buffers used to make a 3-input wired-
AND gate

Figure 6.19 shows the functional diagram and basic
Truth Table of a 74LS365 Hex 3-state buffer IC, in
which all six buffers share a common AND-gated
control line. This IC can be used as six normal buffers

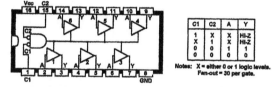

C1	C2	A	Y
1	X	X	HI-Z
X	1	X	HI-Z
0	0	1	1
0	0	0	0

Notes: X = either 0 or 1 logic levels.
Fan-out = 30 per gate.

Figure 6.19   Functional diagram and Truth Table of the 74LS365
Hex 3-state buffer

by grounding its two control pins as shown in *Figure
6.20*, or as a set of six 3-state buffers that are all
switched via one common control signal as shown in
Figure 6.21; AND-type 3-state control can be obtained
by using both 'Control' terminals (pins 1 and 15).

Finally, the 74HC241, 74HC244, and 74LS244 are
'Octal' 3-state Schmitt buffers in which the buffers are
split into two groups of four, with the mode of each
group controlled via a separate input. *Figure 6.22* shows

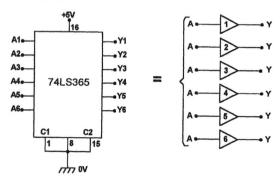

Figure 6.20 Methods of connecting the 74LS365 for use as six normal buffers

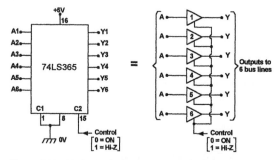

Figure 6.21 Methods of connecting the 74LS365 for use as a 3-state Hex buffer controlled via a single input

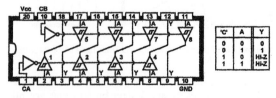

Figure 6.22 Functional diagram and Truth Table of the 74LS244 (or 74HC244) Octal (dual Quad) 3-state Schmitt buffer IC

the functional diagram and Truth Table of the 74LS244 IC (the 74HC244 is similar, but it has Schmitt-type CA and CB inputs); the IC is really a dual Quad device, in which buffers 1–4 are controlled via the CA terminal, and buffers 5–8 are controlled via the CB terminals.

Each of these Quads can be used as a set of simple Schmitt buffers by grounding its control terminal as shown in *Figure 6.23(a)*, or as a ganged set of 3-state Schmitt buffers by using its Control terminal as shown in *Figure 6.23(b)*.

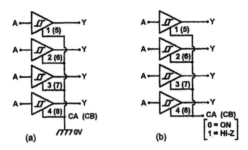

Figure 6.23 The 74LS244 'Quad' buffers can be used as *(a)* simple Schmitt buffers, or as *(b)* 3-state ganged Schmitt buffers

Practical Inverter IC Circuits

The inverter (or NOT gate) is the most basic of all digital logic elements, and is sometimes called an inverting buffer. If you ever need only a few simple inverters, one cheap way to get them is to make them from spare TTL or CMOS NAND or NOR elements, connected as shown in *Figure 6.24*.

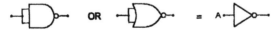

Figure 6.24 Any NAND or NOR gate can be used as an inverting buffer element

Figure 6.25 lists basic details of some popular inverter ICs. When using these ICs, note that all unused inverters must be disabled by tying their inputs to one of the IC's supply lines. In CMOS devices, the unused inputs can be tied directly to either supply line, but in TTL devices it is best (for lowest quiescent current consumption) to tie all unused inputs directly to the 0V rail; if the unused inverter is a 3-state type, it should (if it has independent

Device	Type	Description
7404	TTL	Hex inverter
74HC04	CMOS	Hex inverter
74HCU04	CMOS	Unbuffered Hex inverter
74LS04	LS TTL	Hex inverter
74LS05	LS TTL	Hex inverter with o.c. outputs
7406	TTL	Hex inverter with o.c. outputs
74LS14	LS TTL	Hex Schmitt inverter
74HC14	CMOS	Hex Schmitt inverter
4049UB	CMOS	Unbuffered Hex inverter
74HC4049	CMOS	Hex inverter
4069UB	CMOS	Unbuffered Hex inverter
4502B	CMOS	Hex 3-state inverter
40106B	CMOS	Hex Schmitt inverter
74LS240	LS TTL	Octal 3-state Schmitt inverter

Figure 6.25 Fourteen popular inverter ICs

control) be set into its 'normal' mode via its control input, to minimize current drain.

Dealing now with the individual inverter ICs listed in *Figure 6.25*, *Figure 6.26* shows the functional diagram that is common to the 7404, 74LS04, 74HC04, 74HCU04, and 4069UB Hex inverter ICs. Of these, the 7404 is an ancient Standard TTL IC, the 74LS04 is a modern LS TTL type in which each inverter has a fan-out of 10, the 74HC04 is a fast CMOS type, and the 74HCU04 and 4069UB are unbuffered CMOS types that are suitable for use in linear applications.

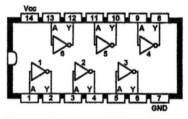

Figure 6.26 Functional diagram of the 7404, 74LS04, 74HC04, 74HCU04, or 4069UB Hex inverter ICs

Figure 6.27 shows the functional diagram that is common to the 74LS05 and 7406 Hex inverters with o.c. outputs; the 74LS05's o.c. outputs can handle maximum outputs of only 5.5 volts, but those of the 7406 can handle up to 30 volts.

Figure 6.28 shows the functional diagram that is common to three of the most useful of all Hex inverter ICs, the

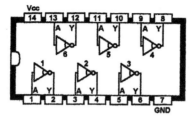

Figure 6.27 Functional diagram of the 74LS05 or 7406 Hex inverters with o.c. outputs

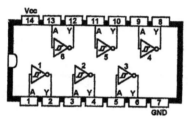

Figure 6.28 Functional diagram of the 74LS14, 74HC14, or 40106B Hex Schmitt inverter ICs

74LS14 TTL and the 74HC14 and 40106B CMOS Schmitt types. In the 74LS14, the output of each Schmitt inverter is in the logic-1 state until the input *rises* to an 'upper threshold' value of 1.6V, at which point the output switches to logic-0 and locks there until the input is *reduced* to a 'lower threshold' value of 0.8V, at which point the output switches and locks into the logic-1 state again, and so on. Thus, a 74LS14 Schmitt inverter can be made to function as a sine-to-square converter by connecting it as shown in *Figure 6.29*, where RV1 is used to set the circuit to its maximum sensitivity point, at which a quiescent voltage of 1.2V is set on the inverter's input.

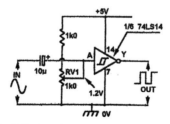

Figure 6.29 A TTL Schmitt inverter sine-to-square converter

Figures 6.30 to *6.32* show more simple applications of the 74LS14 IC. *Figure 6.30* is a practical version of the *Figure 3.6* switch debouncer; it can be activated by a push-button (S1) or toggle (S2) switch, and has an output that goes to logic-1 when the switch is closed. *Figure 6.31* is a modified version of the above circuit, with an added inverter stage, and gives a logic-0 output when S1 is closed. *Figure 6.32* is yet another modification of the basic circuit, and generates a brief logic-1 'switch-on' output pulse when the circuit's supply is first connected.

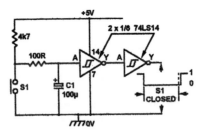

Figure 6.30   TTL switch debouncer, with logic-1 'closed' output

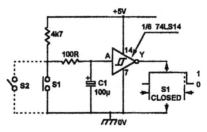

Figure 6.31   TTL switch debouncer, with logic-0 'closed' output

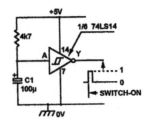

Figure 6.32   TTL supply switch-on pulse generator

Regarding the 74HC14 and 40106B CMOS Schmitt inverters, these have typical upper and lower threshold values equal to 60% and 40% of the supply voltage respectively. A CMOS Schmitt inverter can thus be made to function as a sine-to-square converter by connecting it as shown in *Figure 6.33*, where RV1 is used to set the circuit to its maximum sensitivity point, or as a 'switch-on' pulse generator (which generates a brief logic-1 'switch-on' output pulse when the circuit's supply is first connected) by wiring it as shown in *Figure 6.34*.

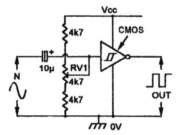

Figure 6.33   CMOS Schmitt inverter sine/square converter

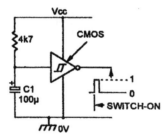

Figure 6.34   CMOS supply switch-on pulse generator

Figure 6.35 shows the functional diagram that is common to the 4049UB and 74HC4049 Hex CMOS inverter ICs. The 4049UB is an unbuffered type, suitable for use in linear applications, and the 74HC4049 is a fast fully buffered general-purpose device.

Figure 6.36 shows the functional diagram and Truth Table of the 4502B. This is a special-purpose 3-state

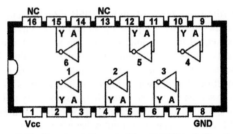

Figure 6.35 Functional diagram of the 4049UB (unbuffered) or 74HC4049 (buffered) Hex inverter IC

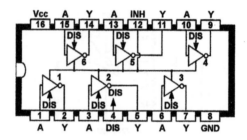

DISable	INHibit	INPUT (A)	OUT (Y)
0	0	0	1
0	0	1	0
0	1	X	0
1	X	X	HI-Z

X = Don't care

Figure 6.36 Functional diagram and Truth Table of the 4502B Hex 3-state inverter with INHIBIT control

Hex inverter in which the outputs of all six inverters can be set to the logic-0 state by driving the INHIBIT (pin-12) terminal high, or can be set to the high-impedance state by driving the DISABLE (pin-4) terminal high; the IC can be used as a conventional Hex inverter by grounding the INHIBIT and DISABLE pins, or as a normal 3-state inverter by grounding the INHIBIT pin and applying the 3-state control to the DISABLE terminal.

Finally, *Figure 6.37* shows the functional diagram and Truth Table of the 74LS240 Octal 3-state Schmitt inverting buffer IC, in which each 'buffer' has a fan-out

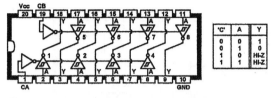

Figure 6.37 Functional diagram and Truth Table of the 74LS240 Octal (dual Quad) 3-state Schmitt inverting buffer IC

of 30. This IC is actually a dual Quad device, in which inverters 1–4 are controlled via the CA terminal, and inverters 5–8 are controlled via the CB terminal. Each of these Quads can be used as a set of normal Schmitt inverters by grounding its control terminal, or as a ganged set of 3-state Schmitt inverters by using its control terminal as shown in the Truth Table.

Practical AND-gate IC Circuits

The output of an AND gate goes high (to logic-1) when all of its inputs (A *and* B *and* C, etc.) are high. *Figure 6.38* lists basic details of several popular AND-gate ICs; of these, the 74LS08, 74HC08, and 4081B (see *Figures 6.39* and *6.40*) are Quad 2-input types, the 74LS11 and 4073B (see *Figures 6.41* and *6.42*) are Triple 3-input types, and the 74LS21 and 4082B (see *Figures 6.43* and *6.44*) are Dual 4-input types.

When using AND-gate ICs, each unwanted gate must be disabled by shorting all of its inputs together and tying

Device	Type	Description
74LS08	LS TTL	Quad 2-input AND gate
74HC08	CMOS	Quad 2-input AND gate
4081B	CMOS	Quad 2-input AND gate
74LS11	LS TTL	Triple 3-input AND gate
4073B	CMOS	Triple 3-input AND gate
74LS21	LS TTL	Dual 4-input AND gate
4082B	CMOS	Dual 4-input AND gate

Figure 6.38 Seven popular AND gate ICs

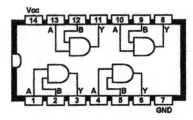

Figure 6.39 Functional diagram of the 74LS08 or 74HC08 Quad 2-input AND gate ICs

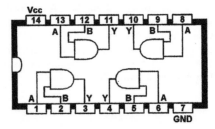

Figure 6.40 Functional diagram of the 4081B Quad 2-input AND gate IC

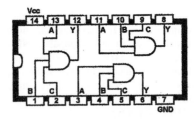

Figure 6.41 Functional diagram of the 74LS11 Triple 3-input AND gate IC

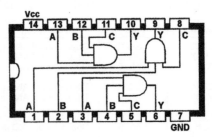

Figure 6.42 Functional diagram of the 4073B Triple 3-input AND gate IC

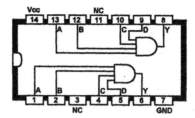

Figure 6.43 Functional diagram of the 74LS21 Dual 4-input AND gate IC

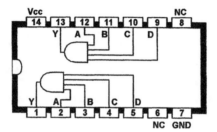

Figure 6.44 Functional diagram of the 4082B Dual 4-input AND gate IC

them to one of the IC's supply lines. In CMOS ICs, the shorted inputs can be wired directly to either supply line, but in TTL ICs the inputs must (to give minimum quiescent current consumption with good stability) be tied to the positive supply rail via a single 1k0 resistor, as shown in *Figure 6.45*; a single resistor can be used as a tie-point for large numbers of unwanted inputs.

Sometimes, when using 3-input or 4-input AND gate ICs, you may not want to use all of a gate's input terminals. In this case, the unwanted inputs can be disabled by either tying them high (directly in CMOS gates, or via a 1k0 resistor in TTL types) or by simply

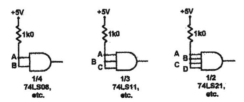

Figure 6.45 Method of disabling unwanted TTL AND gates

shorting them directly to a used input; *Figure 6.46* shows examples of 3-input and 4-input TTL AND gates wired for use as 2-input types. Note that the fan-in of a TTL AND gate is an almost constant '1', irrespective of the number of inputs used. Thus, CMOS or TTL AND gates can be converted into non-inverting buffers by simply shorting all of their inputs together; *Figure 6.47* shows examples of TTL AND gates used as simple buffers.

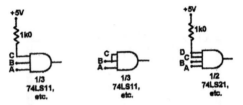

Figure 6.46   Methods of disabling unwanted TTL inputs, to make a 2-input AND gate

Figure 6.47   Methods of using TTL and gates as simple buffers

A useful feature of AND gate ICs is that their gates can be directly cascaded, with the output of one gate feeding directly into one input of another gate, to make compound AND gates with any desired number of inputs. *Figure 6.48*, for example, shows how 2-input AND gates can be cascaded to make 3-input, 4-input, or 5-input AND gates, and *Figure 6.49* shows three 3-input or two 4-input gates cascaded to make a single 7-input AND gate.

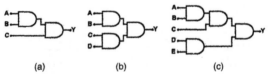

Figure 6.48   Ways of using 2-input AND gates to make *(a)* 3-input, *(b)* 4-input, or *(c)* 5-input AND gates

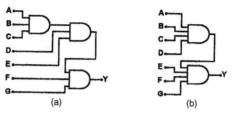

Figure 6.49 Ways of using *(a)* 3-input or *(b)* 4-input AND gates to make a 7-input AND gate

Practical NAND-gate IC Circuits

The output of a NAND gate goes low when all of its inputs (A *and* B, etc.) are high. *Figure 6.50* lists basic details of 13 popular NAND-gate ICs; of these, the 74LS00, 74HC00, and 4011B (see *Figures 6.51* and

Device	Type	Description
4011B	CMOS	Quad 2-input NAND gate
4093B	CMOS	Quad 2-input Schmitt NAND gate
74LS00	LS TTL	Quad 2-input NAND gate
74HC00	CMOS	Quad 2-input NAND gate
74LS132	LS TTL	Quad 2-input Schmitt NAND gate
74HC132	CMOS	Quad 2-input Schmitt NAND gate
74LS10	LS TTL	Triple 3-input NAND gate
4023B	CMOS	Triple 3-input NAND gate
74LS20	LS TTL	Dual 4-input NAND gate
4012B	CMOS	Dual 4-input NAND gate
74LS30	LS TTL	8-input NAND gate
4068B	CMOS	8-input NAND gate
74HC133	CMOS	13-input NAND gate

Figure 6.50 Thirteen popular NAND gate ICs

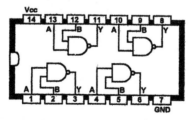

Figure 6.51 Functional diagram of the 74LS00 or 74HC00 Quad 2-input NAND gate IC

6.52) are standard Quad 2-input types, and the 4093B, 74LS132 and 74HC132 (see *Figures 6.53* and *6.54*) are Schmitt Quad 2-input types. The 74LS10 and 4023B

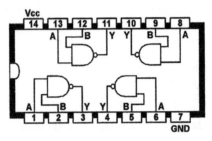

Figure 6.52   Functional diagram of the 4011B Quad 2-input NAND gate IC

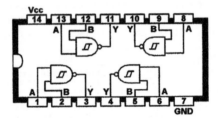

Figure 6.53   Functional diagram of the 4093B Quad 2-input Schmitt NAND gate IC

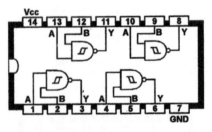

Figure 6.54   Functional diagram of the 74LS132 or 74HC132 Quad 2-input Schmitt NAND gate IC

(see *Figures 6.55* and *6.56*) are Triple 3-input standard types, the 74LS20 and 4012B (see *Figures 6.57* and *6.58*) are Dual 4-input standard types, the 74LS30 and 4068B (see *Figures 6.59* and *6.60*) are 8-input standard types, and the 74HC133 is a 13-input standard type.

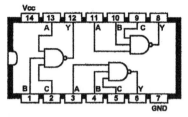

Figure 6.55 Functional diagram of the 74LS10 Triple 3-input NAND gate IC

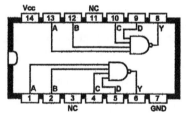

Figure 6.56 Functional diagram of the 4023B Triple 3-input NAND gate IC

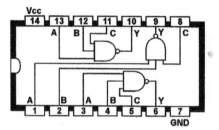

Figure 6.57 Functional diagram of the 74LS20 Dual 4-input NAND gate IC

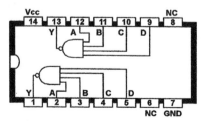

Figure 6.58 Functional diagram of the 4012B Dual 4-input NAND gate IC

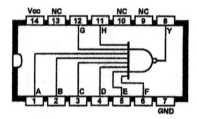

Figure 6.59   Functional diagram of the 74LS30 8-input NAND gate IC

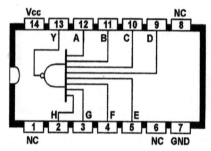

Figure 6.60   Functional diagram of the 4068B 8-input NAND gate IC

When using NAND-gate ICs, each unwanted gate should be disabled by shorting all of its inputs together and tying them to one of the IC's supply lines. In CMOS ICs, the shorted inputs can be wired directly to either supply line, but in TTL ICs the inputs must (to give minimum quiescent current consumption with good stability) be tied directly to the 0V rail, as shown in *Figure 6.61*.

Sometimes, when using NAND gate ICs, you may not want to use all of a gate's input terminals. In this case, the unwanted inputs can be disabled by either tying them high (directly in CMOS gates, or via a 1k0 resistor

Figure 6.61   Methods of disabling unwanted TTL NAND gates

in TTL types) or by simply shorting them directly to a used input; *Figure 6.62* shows examples of a 3-input TTL NAND gate wired for use as a 2-input type. Note that the fan-in of a TTL NAND gate is an almost constant '1', irrespective of the number of inputs used. Thus, CMOS or TTL NAND gates can be converted into simple inverters by simply shorting all of their inputs together; *Figure 6.63* shows examples of TTL NAND gates used as inverters. Also note that NAND gates are fairly versatile elements, as demonstrated in *Figure 6.64*, which shows ways of using 2-input elements to make a 2-input or 3-input AND gate or a 3-input NAND gate.

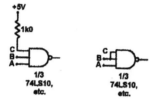

Figure 6.62 Basic method of disabling unwanted TTL NAND gate inputs

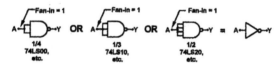

Figure 6.63 Methods of using TTL NAND gates as simple inverters

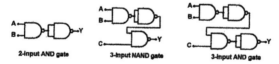

Figure 6.64 Ways of using 2-input NAND gates to make various AND and NAND gates

Practical OR-Gate IC Circuits

The output of an OR gate goes high when any of its inputs (A *or* B, etc.) go high. The simplest way to make an OR gate is via a number of diodes and a single resistor, as shown (for example) in the 3-input OR gate of *Figure 6.65*. The diode OR gate is reasonably fast, very cost-effective, and can readily be expanded to accept any number of inputs by simply adding one more diode to the circuit for each new input.

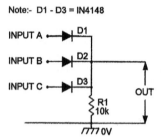

Note:- D1 - D3 = IN4148

Figure 6.65   Simple 3-input diode OR gate

Relatively few dedicated OR-gate ICs are available; *Figure 6.66* lists basic details of the six most popular OR-gate ICs; the 74LS32, 74HC32, and 4071B (see *Figures 6.67* and *6.68*) are Quad 2-input types, the 4075B and 74HC4075 (see *Figure 6.69*) are Triple 3-input types, and the 4072B (see *Figure 6.70*) is a Dual 4-input type.

Device	Type	Description
74LS32	LS TTL	Quad 2-input OR gate
74HC32	CMOS	Quad 2-input OR gate
4071B	CMOS	Quad 2-input OR gate
4075B	CMOS	Triple 3-input OR gate
74HC4075	CMOS	Triple 3-input OR gate
4072B	CMOS	Dual 4-input OR gate

Figure 6.66   Six popular OR gate ICs

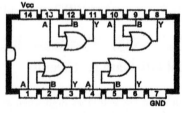

Figure 6.67   Functional diagram of the 74LS32 or 74HC32 Quad 2-input OR gate IC

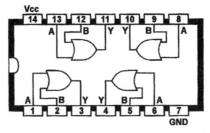

Figure 6.68 Functional diagram of the 4071B Quad 2-input OR gate IC

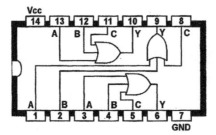

Figure 6.69 Functional diagram of the 4075B or 74HC4075 Triple 3-input OR gate IC

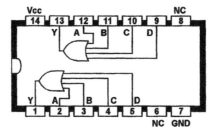

Figure 6.70 Functional diagram of the 4072B Dual 4-input OR gate IC

When using OR-gate ICs, each unwanted gate should be disabled by shorting all of its inputs together and tying them to one of the IC's supply lines. In CMOS ICs, the shorted inputs can be wired directly to either supply line, but in TTL ICs the inputs must (to give minimum quiescent current consumption with good stability) be tied high via a 1k0 resistor, as shown in *Figure 6.71*.

Figure 6.71 Method of disabling a TTL OR gate

Note that the fan-in of a TTL NOR gate is directly proportional to the number of inputs used, at a fan-in rate of 1-per-input, and that a TTL 2-input OR gate can be made to act as a simple non-inverting buffer by either tying one input to ground or by tying both inputs together, as shown in *Figure 6.72*, but that the buffer has

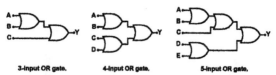

Figure 6.72 Ways of using a TTL OR gate as a simple buffer

a fan-in of 1 in the former case, and a fan-in of 2 in the latter. Also note that OR gates can be directly cascaded to make a compound OR gate with any desired number of inputs; *Figure 6.73*, for example, shows ways of

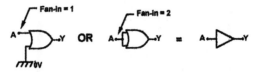

3-input OR gate.   4-input OR gate.   5-input OR gate.

Figure 6.73 Ways of cascading 2-input OR gates to give up to 5 inputs

cascading 2-input elements to make OR gates with 3, 4, or 5 inputs, and *Figure 6.74* shows a 3-input OR element and a 3-input diode OR gate cascaded to make a compound 5-input OR gate.

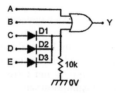

Figure 6.74 Example of a compound 5-input OR gate circuit

Practical NOR-gate IC Circuits

The output of a NOR gate goes low when any of its inputs (A *or* B, etc.) go high. One easy way to make a NOR gate is to combine a basic diode OR gate with a transistor or IC inverter stage, as shown in the 3-input NOR gate circuit of *Figure 6.75*. NOR gates of this type are reasonably fast and cost-effective and can easily be expanded to accept any desired number of inputs by simply adding one new diode for each new input.

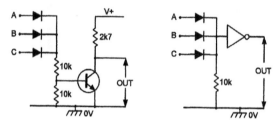

Figure 6.75 A simple NOR gate can be made by feeding the output of a diode OR gate through a transistor or IC inverter stage

Figure 6.76 lists basic details of the ten most popular NOR-gate ICs; of these, the 74LS02, 74HC02, 4001B and the unbuffered 4001UB (see *Figures 6.77* and *6.78*) are all Quad 2-input types. The 4025B and the 74LS27 and 74HC27 (see *Figures 6.79* and *6.80*) are Triple 3-input types, and the 4002B and 74LS260 (see *Figures 6.81* and *6.82*) are Dual 4-input and 5-input types

Device	Type	Description
74LS02	LS TTL	Quad 2-input NOR gate
74HC02	CMOS	Quad 2-input NOR gate
4001B	CMOS	Quad 2-input NOR gate
4001UB	CMOS	Unbuffered Quad 2-input NOR gate
4002B	CMOS	Dual 4-input NOR gate
74LS260	LS TTL	Dual 5-input NOR gate
4025B	CMOS	Triple 3-input NOR gate
74LS27	LS TTL	Triple 3-input NOR gate
74HC27	CMOS	Triple 3-input NOR gate
4078B	CMOS	8-input NOR gate

Figure 6.76 Ten popular NOR gate ICs

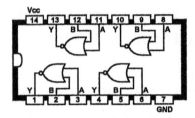

Figure 6.77 Functional diagram of the 74LS02 or 74HC02 Quad 2-input NOR gate IC

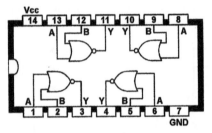

Figure 6.78 Functional diagram of the 4001B or 4001UB Quad 2-input NOR gate IC

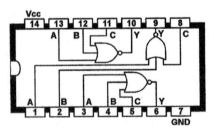

Figure 6.79 Functional diagram of the 4025B Triple 3-input NOR gate IC

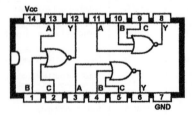

Figure 6.80 Functional diagram of the 74LS27 or 74HC27 Triple 3-input NOR gate IC

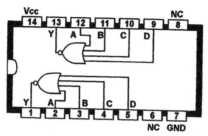

Figure 6.81 Functional diagram of the 4002B Dual 4-input NOR gate IC

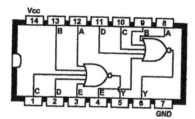

Figure 6.82 Functional diagram of the 74LS260 Dual 5-input NOR gate IC

respectively. The 4078B (*Figure 6.83*) is an 8-input NOR gate IC.

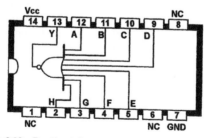

Figure 6.83 Functional diagram of the 4078B 8-input NOR gate IC

When using NOR-gate ICs, each unwanted gate should be disabled by shorting all of its inputs together and tying them to one of the IC's supply lines. In CMOS ICs, the shorted inputs can be wired directly to either supply line, but in TTL ICs the inputs must (to give minimum quiescent current consumption with good stability) be tied directly to the 0V rail, as shown in *Figure 6.84.*

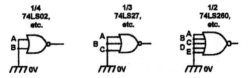

Figure 6.84 Ways of disabling TTL NOR gates

Sometimes, when using NOR gate ICs, you may not want to use all of a gate's input terminals. In this case, the unwanted inputs are best disabled by shorting them directly to the 0V rail, as shown in the examples of *Figure 6.85.* A NOR gate can be made to act as a simple

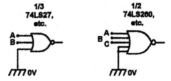

Figure 6.85 Basic ways of disabling unwanted TTL NOR gate inputs

inverter by either shorting all of its inputs together or by grounding all but one of its inputs; note, however, that the fan-in of a TTL NOR gate is directly proportional to the number of inputs used, so the first method is thus (theoretically) the best, since it offers the lowest fan-in value, as shown in *Figure 6.86.*

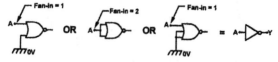

Figure 6.86 Ways of using a TTL NOR gate as a simple inverter

NOR gates are fairly versatile devices. The effective number of inputs of a NOR gate can be increased by applying the extra inputs via a diode or IC OR gate, as shown in *Figure 6.87. Figure 6.88* shows various ways of using 2-input elements to make 2-input or 3-input OR gates or a 3-input NOR gate. Note that a NOR gate can

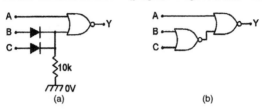

Figure 6.87 Ways of increasing the effective number of inputs of a NOR gate

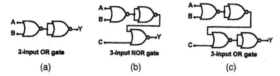

Figure 6.88 Ways of using 2-input NOR gates to make various OR and NOR gates

be converted into an AND gate by simply inverting all of its inputs, and *Figure 6.89* shows how three 2-input NOR gates can be used to make a single AND gate.

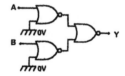

Figure 6.89 NOR gates used to make a 2-input AND gate

Practical EX-OR and EX-NOR IC Circuits

The output of an EX-OR gate goes high only when its two inputs are at different logic levels. The most widely used EX-OR gate ICs are the 74LS86 TTL and the 74HC86 and 4070B CMOS Quad types (see *Figures 6.90* and *6.91*). If one or more of a CMOS EX-OR IC's gates are unwanted, it can be disabled by simply grounding both inputs. In the case of a TTL EX-OR IC, unwanted gates are best disabled by grounding one input terminal and tying the other high via a 1k0

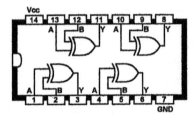

Figure 6.90   Functional diagram of the 74LS86 or 74HC86 Quad EX-OR gate IC

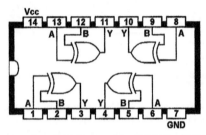

Figure 6.91   Functional diagram of the 4070B Quad EX-OR gate IC

resistor, as shown in *Figure 6.92*, as this results in minimum quiescent current consumption; alternatively, if current drain is not important, both inputs can simply be tied to ground as shown.

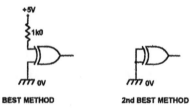

Figure 6.92   Two ways of disabling unwanted TTL EX-OR gates (see text)

EX-OR gates are quite versatile. An EX-OR gate can be made to act as a non-inverting buffer by simply grounding its unused input, as shown in *Figure 6.93*(a), or as an inverting buffer by tying the unused input high (via a 1k0 resistor in TTL types), as shown in *Figure 6.93*(b). Thus, two EX-OR gates can be used to make a single EX-NOR gate by connecting them as shown in *Figure 6.94*, where

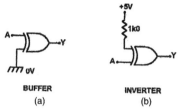

BUFFER
(a)

INVERTER
(b)

Figure 6.93 Ways of using TTL EX-OR gates as *(a)* buffers or *(b)* inverters

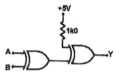

Figure 6.94 An EX-NOR gate made from two TTL EX-OR gates

the right-hand gate is used to invert the output of the left-hand gate; note that Quad EX-NOR CMOS ICs are also available in dedicated IC forms as the 4077B and the 74HC266 (see *Figures 6.95* and *6.96*), but that the latter IC's gates have open-drain outputs.

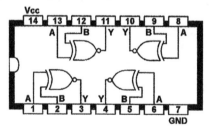

Figure 6.95 Functional diagram of the 4077B Quad EX-NOR gate IC

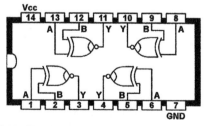

Figure 6.96 Functional diagram of the 74HC266 Quad EX-NOR gate IC with open-drain (o.d.) outputs

Figures 6.97 and *6.98* show two other useful EX-OR gate applications. In *Figure 6.97*, four EX-OR gates are fed with a common control signal that enables a 4-bit 'ABCD' input code to be presented in the form of either a true (direct) or complement (inverted) ABCD output, thus making the 4-bit True/Complement outputs available via five (rather than eight) terminals. The circuit in *Figure 6.98* simply compares the logic states of two 4-bit words and gives a logic-0 output if the two words are identical, and a logic-1 output if they differ.

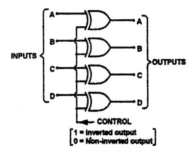

Figure 6.97　4-bit True/Complement generator

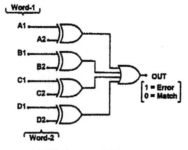

Figure 6.98　4-bit logic state comparator

One of the most important applications of the EX-OR gate is as a binary adder. *Figure 6.99* lists the basic rules

```
0 + 0 = 0
0 + 1 = 1
1 + 0 = 1
1 + 1 = 0, plus carry 1.
```

Figure 6.99　Basic rules of binary addition

of binary addition, and *Figure 6.100* shows how an EX-OR and an AND gate can be used to make a practical 'half-adder' circuit that can add two binary inputs together and generate SUM and CARRY outputs. The circuit is called a 'half-adder' because it can perform only a very primitive form of addition that does not enable it to accept a 'carry' input from a previous addition stage. A 'full-adder' is a far more useful circuit that can accept a 'carry' input, perform 2-bit binary addition, and generate a 'carry' output; such circuits are fully cascadable, enabling groups of circuits to perform binary addition on digital numbers of a desired 'bit' size. *Figure 6.101* shows one way of building a 2-bit full-adder circuit, using three EX-OR gates, two AND gates, and an OR gate. In practice, 4-bit full-adders are readily available in the forms of the 74LS283 and 4008B ICs.

A	B	SUM	CARRY
0	0	0	0
0	1	1	0
1	0	1	0
1	1	0	1

Figure 6.100 Binary half-adder circuit

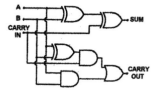

A	B	CARRY IN	SUM	CARRY OUT
0	0	0	0	0
0	1	0	1	0
1	0	0	1	0
1	1	0	0	1
0	0	1	1	0
0	1	1	0	1
1	0	1	0	1
1	1	1	1	1

Figure 6.101 Full-adder circuit and Truth Table

Mixed Gate ICs

The '74' and '4000B' ranges of ICs include a number of 'mixed gate' types that contain gates of more than one type, wired together for use in special applications. The best known of these are 'AND-OR-INVERT' ('AOI') gates, and examples of these are shown in *Figures 6.102* and *6.103*. It is unlikely that you will ever need to use an AOI gate, but you may find it useful to learn some of the

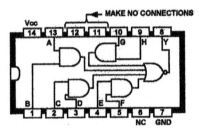

Figure 6.102 Functional diagram of the 7454 4-wide 2-input AND-OR-INVERT (AOI) gate IC

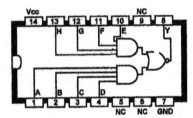

Figure 6.103 Functional diagram of the 74LS55 2-wide 4-input AND-OR-INVERT gate IC

jargon that is associated with them. Looking first at the 7454 IC of *Figure 6.102*, note that it has input connections to four 2-input AND gates, and this accounts for the '4-wide 2-input AND-' part of the IC's title; also note that the outputs of these four AND gates are wired to the inputs of a NOR gate and thence made externally available via pin-8; now, a NOR gate is simply an OR gate with an inverted output, and this fact accounts for the 'OR-INVERT' part of the ICs title. Thus, an 'AND-OR-INVERT' gate is simply an 'AND-NOR gate' with a rather flowery title. The action of the 7454 IC is such that its output is normally high, and goes low only when both inputs to one or more AND gates are high.

Turning next to the 74LS55 IC of *Figure 6.103*, note that this has input connections to two 4-input AND gates, and this accounts for the IC's '2-wide 4-input AND-OR-INVERT' title; the IC's action is such that its output is normally high, and goes low only when all inputs to at least one AND gate are high.

Returning briefly to the 7454 IC, note that this is a Standard-TTL device, and its data sheet carries a very sinister warning that NO EXTERNAL CONNECTION must be made to pins 11 and 12. Also note that the 'LS' version of this device (the 74LS54) has two of its four AND gates configured as 2-input types and two configured as 3-input types, and is sometimes known as a '4-wide, 2–3–3–2-input AND-OR-INVERT gate', and has pin connections, etc., that differ from those shown in *Figure 6.102.*

A 'Programmable' Logic IC

Most logic ICs are dedicated devices that contain a number of 'fixed' gates. One very useful exception is the CMOS IC known as the 4048B 'programmable' 8-input multifunction gate (see *Figure 6.104*). This modestly priced (but hard to find) 16-pin IC has two groups of four input terminals, plus an 'expansion' input terminal, and is provided with four control (K) pins which let the user select the mode of logic operation, plus a 'J' output terminal that enables 4048B ICs to be cascaded, so that two of them make a 16-input gate, or four make a 32-input gate, and so on.

Control input pin Kd (pin-2) enables the user to select either normal or high-impedance 3-state output operation. The remaining three binary control inputs (Ka, Kb, and Kc) enable any one of eight different logic functions to be selected, as shown in the table of *Figure 6.105*, which also shows how to connect unwanted inputs in each mode of operation. Thus, to make the 4048B act as a normal 6-input OR gate, connect the two unwanted inputs to ground (logic-0), and connect Ka and Kb to ground and Kc and Kd to the positive supply rail. The EXPAND input (pin-15) is normally tied to ground.

Eight different logic functions are available from the 4048B, as shown in *Figure 6.106*. Operation in the AND, OR, NAND and NOR modes is quite conventional, but operation in the remaining four modes (OR/AND, OR/NAND, AND/OR, and AND/NOR) is less self-

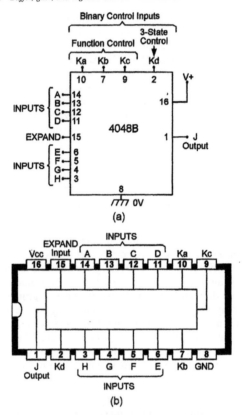

Figure 6.104 Functional diagram *(a)* and outline *(b)* of the 4048B multifunction expandable 8-input gate IC

OUTPUT FUNCTION	Ka	Kb	Kc	UNUSED INPUTS
NOR	0	0	0	0
OR	0	0	1	0
OR/AND	0	1	0	0
OR/NAND	0	1	1	0
AND	1	0	0	1
NAND	1	0	1	1
AND/NOR	1	1	0	1
AND/OR	1	1	1	1
Kd = 1 for Normal action = 0 for High-Z outputs EXPAND input = 0				

Figure 6.105 Function control table of the 4048B 8-input gate IC

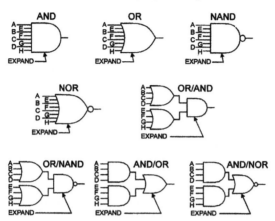

Figure 6.106 The eight basic logic functions of the 4048B IC

evident. In the latter cases the inputs are split into two groups of four, each of which provides the 'first' part of the logic function, but the *pair* of groups provide the 'second' part of the logic function. Thus, in the OR/AND mode, the IC gives a high output only if at least one input is present in the A-to-D group at the same time as at least one input is present in the E-to-H group, etc.

The EXPAND input terminal of the 4048B enables ICs to be cascaded; thus, two ICs can (for example) be made to act as a 16-input gate by feeding the output of one IC into the EXPAND terminal of the other. Note when using 'expanded' logic that the input logic feeding the 'expand' terminal is not necessarily the same as the overall logic that is required. Thus, an OR 'expand' input is needed for expanded NOR or OR operation, a NAND 'expand' for AND and NAND operation, a NOR 'expand' for OR/AND or OR/NAND operation, and an AND 'expand' for AND/OR and AND/NOR operation.

'Majority Logic' Circuits

One little known but useful type of logic system is 'majority' logic, in which the logic device has an odd (3,

5, 7, etc.) number of inputs and gives an active output only when the majority of these inputs are high, irrespective of *which* inputs are high. This type of logic is useful in pseudo-intelligent alarms and robotic devices, etc., and may (for example) sound an alarm bell only if at least two of three detectors indicate a 'fault' condition, or enable a robot to move only if there is more stimulus to move than there is to stand still, etc.

The best known CMOS majority logic IC is the 4530B Dual 5-input unit (*Figure 6.107*), each half of which contains a 5-input majority logic element with its output feeding to one input of an EX-NOR gate that has its other input (W) externally available, enabling it to be wired as either an inverting or non-inverting stage. Thus, when 'W' is tied to logic-1 the EX-NOR stage gives non-inverting action and the element's output goes high only when the majority of inputs are high, but when 'W' is tied to logic-0 the EX-NOR stage gives an inverting action and the element's output goes high when the majority of inputs are low. Note that the effective number of inputs of a 4530B element can be reduced by wiring half of the unwanted inputs to logic-1 and the other half to logic-0, as in *Figure 6.108(a)*. Alternatively, the effective number of inputs can be increased by cascading elements, with the output of one element feeding one input of the cascaded element, as in *Figure 6.108(b)*.

In practice, the 4530B IC is often hard to find. In this case a majority logic circuit can easily be built by wiring a 3140 CMOS op-amp in the basic configuration of *Figure 6.109*, which shows a 5-input circuit. Here, the

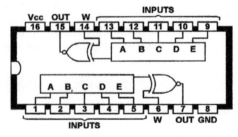

Figure 6.107 Functional diagram of the 4530B Dual 5-input majority-logic gate IC

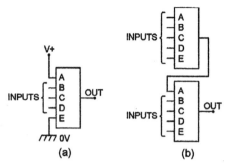

Figure 6.108 The number of effective inputs of a majority-logic circuit can be *(a)* decreased or *(b)* increased easily

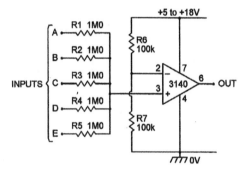

Figure 6.109 Simple 5-input op-amp majority-logic gate

op-amp functions as a voltage comparator in which R6–R7 applies half of the supply voltage to pin-2 of the op-amp, and the five input resistors (which must each be connected to either ground or the positive supply rail) form a potential divider that applies a fraction of the supply voltage to pin-3. This pin-3 voltage is lower than that of pin-2 if the majority of inputs are low, but greater than that of pin-2 if the majority of inputs are high; under this latter condition the op-amp output switches high, and this gives 'majority logic' action.

Note that if 5% resistors are used, the *Figure 6.109* circuit can be given any odd number of inputs up to a maximum of 11 by simply adding one more 1M0 resistor for each new input. The output of this circuit switches almost fully to zero volts when the output is low, but only rises to within a couple of volts of the positive supply rail when the output is high. In most applications this defect is of little importance; it does mean, however, that elements cannot be cascaded to increase the total number of inputs. This defect can be overcome by using the alternative configuration of *Figure 6.110*, in which the output is inverted and level shifted by Q1, and the inputs to the op-amp are transposed. The output of this circuit switches to within 50mV of either supply rail, enabling units to be cascaded without limit.

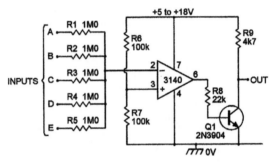

Figure 6.110 Compound 5-input op-amp majority-logic gate

Assertion-level Logic Notion

The reader is already familiar with the fact that an AND or OR gate has an 'active-high' output (i.e. the output of an OR gate goes high when any input is high, etc.), and with the MIL/ANSI convention that the addition of a little circle to an OR gate output (etc.) implies that the gate has an 'active low' output, as shown in *Figure 6.111*. Technically, the presence or absence of this 'little

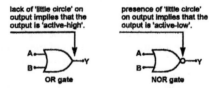

Figure 6.111 Basic principle of 'assertion-level' logic notation

circle' is known as 'assertion-level logic notation', and it can be legitimately applied both to the input or the output of a logic symbol.

Thus, the crude 'gated pulse generator' symbol of *Figure 6.112(a)* implies that the pulse generator is gated on by a HIGH input signal, but the modified symbol of *(b)* – in which a little circle is added to the generator's input – implies that this generator is gated on by a LOW input signal, etc.; note that in practice the *(a)* generator can be made to give the same action as that of *(b)* by simply inserting an inverter stage between the IN terminal and the input of the generator, as shown in *(c)*,

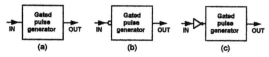

Figure 6.112   Pulse generators that can be gated-on by *(a)* logic-1 and *(b)* and *(c)* logic-0 inputs

so that a LOW input forces the generator's input HIGH and gates it on. This system of assertion-level logic notation is in fact very widely used in electronic logic symbology; some examples of its use are shown in *Figure 6.113*, which deals with mixed logic equivalents.

Mixed Logic Equivalents

When assertion-level logic notation is applied to a simple 2-input AND or OR gate, it can be quickly seen that the gate has four possible input assertion-level sets, i.e. both inputs active-HIGH, or both active-LOW, or one active-HIGH and one active-LOW, or vice versa. Similarly, the gate's output has two possible assertion-levels (active-HIGH or active-LOW). Thus, a 2-input AND or OR gate has a total of eight possible input/output assertion-levels. If Truth Tables are drawn up for all 16 possible AND and OR gate variations, it becomes apparent that each AND gate variation has a mixed-logic OR gate equivalent, and vice versa, as shown in *Figure 6.113*. Note in particular that a normal

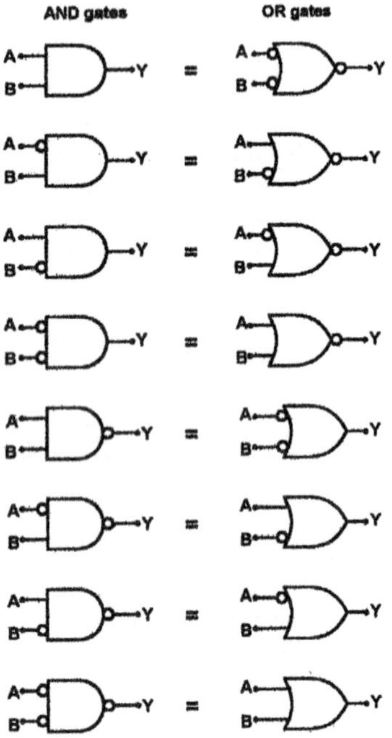

AND gates   **OR gates**   Truth Tables

A	B	Y
0	0	0
0	1	0
1	0	0
1	1	1
0	0	0
0	1	1
1	0	0
1	1	0
0	0	0
0	1	0
1	0	1
1	1	0
0	0	1
0	1	0
1	0	0
1	1	0
0	0	1
0	1	1
1	0	1
1	1	0
0	0	1
0	1	0
1	0	1
1	1	1
0	0	1
0	1	1
1	0	0
1	1	1
0	0	0
0	1	1
1	0	1
1	1	1

Figure 6.113   Mixed logic equivalents

AND gate can be simulated by a NOR gate with both inputs inverted, and that a normal OR gate can be simulated by a NAND gate with both inputs inverted, etc.

Digital 'Transmission' Gates

Most logic gate circuits presented in this chapter show the gates used as simple logic state detectors. 2-input AND, NAND, OR and NOR gates can, however, also be used as digital 'transmission' gates which pass a digital input signal only when they are 'opened' by an appropriate control signal or logic level. Transmission gates are available in four basic types, and the logic symbols of these are shown, together with their logic gate equivalents, in *Figure 6.114*. The basic transmission

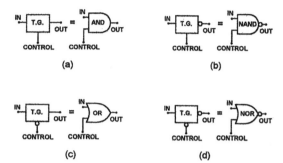

Figure 6.114 Four basic types of digital transmission gate, with their 2-input logic-gate equivalents

gate *(a)* gives a non-inverted output and can be opened by a logic-1 control signal, and can be simulated by a 2-input AND gate; this transmission gate can be made to give an inverted output, as in *(b)*, by using a NAND gate instead of an AND gate. Another variation of the basic transmission gate is shown in *(c)*; it gives a non-inverted output but is opened by a logic-0 control signal, and can be simulated by a 2-input OR gate; this gate can be made to give an inverted output, as in *(d)*, by using a NOR gate instead of an OR gate. Finally, to conclude this chapter, *Figures 6.115* to *6.118* show the precise

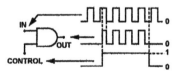

Figure 6.115 AND-type transmission gate is opened by logic-1 control and gives normally-low non-inverted output

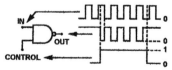

Figure 6.116 NAND-type transmission gate is opened by logic-1 control and gives normally-high inverted output

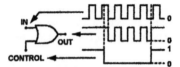

Figure 6.117 OR-type transmission gate is opened by logic-0 control and gives normally-high non-inverted output

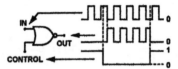

Figure 6.118 NOR-type transmission gate is opened by logic-0 control and gives normally-low inverted output

relationships between the input, output, and control signals of each of these four types of transmission gate.

7 Bilateral Switches and Selectors

A CMOS bilateral switch or transmission gate can be regarded as a near-perfect single-pole single-throw (SPST) electronic switch that can pass analogue or digital signals in either direction and can be turned on (closed) or off (opened) via a logic-1 or logic-0 signal applied to a high-impedance control terminal. Such switches have a near-infinite OFF impedance, and an ON impedance of a few tens or hundreds of ohms.

Practical CMOS bilateral switches have useful operating frequencies ranging up to a maximum of 120MHz, and have many practical uses. They can, for example, be used to replace mechanical switches in signal-carrying applications, the bilateral switch being DC controlled and placed directly on the PCB, exactly where it is needed, thus eliminating the problems of signal radiation and interaction that normally occur when such signals are mechanically switched via lengthy cables. CMOS bilateral switches can also be used in such diverse applications as signal gating, multiplexing, A-to-D and D-to-A conversion, digital control of frequency, impedance and signal gain, the synthesis of multigang potentiometers and capacitors, and the implementation of sample-and-hold circuits, etc. Practical examples of most of these applications are shown later in this chapter.

Several types of CMOS multiple bilateral switch IC are available, in both standard '4000'-series and fast '74HC'-series forms. These range from simple types housing four independently accessible SPST bilateral switches, to fairly complex types housing an array of bilateral switches and logic networks arranged in the form of two independently accessible single-pole 8-way bilateral switches or multiplexers/demultiplexers. Before looking at practical ICs, however, let's look at the basic operating principles and terminology, etc., of the bilateral switch.

Basic Principles

Figure 7.1 shows *(a)* the basic circuit and *(b)* the equivalent circuit of a simple '4000'-series CMOS bilateral switch. Here, an n-type and a p-type MOSFET are effectively wired in inverse parallel (drain-to-source and source-to-drain), but have their gates biased in anti-phase from the control terminal via a pair of inverters. When the control signal is at the logic-0 level, the gate of Q2 is driven to V_{DD} and the gate of Q1 is driven to V_{SS}; under this condition both MOSFETs are cut off, and an effective open circuit exists between the X and Y points of the circuit. When, on the other hand, the control signal is set at the logic-1 level, the gate of Q2 is driven to V_{SS} and the gate of Q1 is driven to V_{DD}, and under this condition both MOSFETs are driven to saturation, and a low impedance exists between the X and Y points.

Note that, when Q1 and Q2 are saturated, signal currents can flow in either direction between the X and Y terminals, provided that the signal voltages are within the V_{SS}-to-V_{DD} limits. Each of the X and Y terminals can thus be used as either an IN or OUT terminal.

In practice, Q1 and Q2 exhibit a finite resistance (R_{ON}) when they are saturated, and in this simple '4000'-series circuit the actual value of R_{ON} may vary from 300R to 1k5, depending on the magnitude of the V_{SS}-to-V_{DD} supply voltage and on the magnitude and polarity of the actual input signal. The simple bilateral switch can thus be represented by the equivalent circuit of *Figure 7.1(b)*.

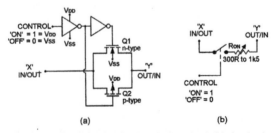

Figure 7.1   Basic circuit *(a)* and equivalent circuit *(b)* of a simple CMOS bilateral switch

Figure 7.2 shows an improved version of the '4000'-series CMOS bilateral switch, together with its equivalent circuit. This circuit is similar to the above, except for the addition of a second bilateral switch (Q3–Q4) that is wired in series with Q5, with the 'well' of Q1 tied to the Q5 drain. These modifications cause Q1's well to switch to V_{SS} when the Q1–Q2 bilateral switch is off, but to be tied to the X input terminal when the switch is on. This reduces the ON resistance of the Q1–Q2 bilateral switch to about 90R and virtually eliminates variations in its value with varying signal voltages, etc., as indicated by the equivalent circuit of *Figure 7.2(b)*. The only disadvantage of the *Figure 7.2* circuit is that it has a slightly lower leakage resistance than *Figure 7.1*.

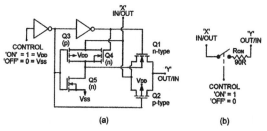

Figure 7.2 Basic circuit *(a)* and equivalent circuit *(b)* of an improved CMOS bilateral switch

Note that '74HC'-series bilateral switch ICs use similar forms of construction to those described above, but are designed around silicon-oxide (rather than Metal-Oxide) MOSFETs, are usually designed to give optimum performance when powered from a 10 volt supply, and have a far lower ON resistance and a far higher maximum operating frequency than '4000'-series types.

Switch Biasing

A CMOS bilateral switch can be used to switch or gate either digital or analogue signals, but must be correctly biased to suit the type of signal being controlled. *Figure*

7.3 shows the basic ways of activating and biasing the bilateral switch. *Figure 7.3(a)* shows that the switch can be turned ON (closed) by taking the control terminal to V_{DD}, or turned OFF (open) by taking the control terminal to V_{SS}.

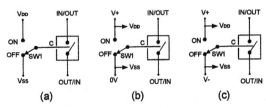

Figure 7.3   Basic method of turning the bilateral switch on and off *(a)*, and power supply connections for use with *(b)* digital and *(c)* analogue IN/OUT signals

In digital signal switching applications (*Figure 7.3(b)*) the bilateral switch can be used with a single-ended supply, with V_{SS} taken to zero volts and V_{DD} at a positive value equal to (or greater than) that of the digital signal (up to a maximum of +15V in a '4000'- series IC). In analogue switching applications (*Figure 7.3(c)*) a split supply (either true or effective) must be used, so that the signal is held at a mean value of 'zero' volts; the positive supply rail goes to V_{DD}, which must be greater than the peak positive voltage value of the input signal, and the negative rail goes to V_{SS} and must be greater than the peak negative value of the input signal; the supply values are limited to plus or minus 7.5V maximum in a '4000'-series IC. Typically, the bilateral switch introduces less than 0.5% of signal distortion when used in the analogue mode.

Logic Level Conversion

Note from the above description of the analogue system that, if a split supply is used, the switch control signal must switch to the positive rail to turn the bilateral switch on, and to the negative rail to turn the switch off. This arrangement is inconvenient in some applications, and some CMOS bilateral switch ICs (notably the

4051B to 4053B family) have built-in 'logic level conversion' circuitry which enables the bilateral switches to be controlled by a digital signal that switches between zero (V_{SS}) and positive (V_{DD}) volts, while still using split supplies to give correct biasing for analogue operation, as shown in *Figure 7.4*.

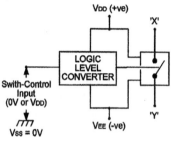

Figure 7.4 Some ICs feature internal 'logic level conversion', enabling an analogue switch to be controlled via a single-ended input

Multiplexing/demultiplexing

A multiplexer can be simply described as a device that enables information from a number (n) of independent data lines to be individually selected and applied to a single DATA or output line, and acts rather like a single-pole multi-way electronic data selection switch. A demultiplexer is the opposite of a multiplexer, and enables the input from a single DATA line to be distributed to any selected one of a number (n) of independent output lines, and acts like a single-pole multi-way data distribution switch.

Figure 7.5 shows how a 4-way demultiplexer (represented by a 4-way switch) can be used to control (turn on or off) four LEDs down a single DATA line. Assume here that the demultiplex driver continuously sequences the demultiplexer through the 1–2–3–4 cycle at a fairly rapid rate, and is synchronized to the 1–2–3–4 segments of the DATA line. Thus, in each cycle, in the '1' period LED1 is off; in the '2' period LED2 is on; in

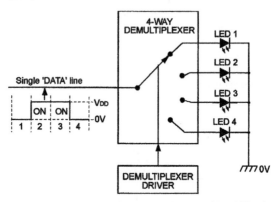

Figure 7.5   A 4-way demultiplexer used to control four LEDs via a single DATA line

the '3' period LED3 is on; and in the '4' period LED4 is off. The state of each of the four LEDs is thus controlled via the logic bit of the single (sequentially time/shared) DATA line.

Figure 7.6 shows how a 4-way multiplexer can be used to feed three independent 'voice' signals down a single cable, and how a demultiplexer can be used to convert the signal back into three independent voice signals at the other end. In practice, each 'sample' period of the DATA line must be short relative to the period of the highest voice frequency; the 'sync' signal is used to synchronize the timing at the two ends of the DATA line.

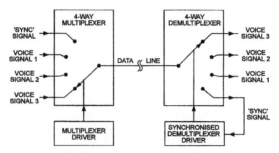

Figure 7.6   A 4-way multiplexer/demultiplexer combination used to feed three independent voice signals through a single DATA line

Note from the above descriptions that a CMOS n-channel multiplexer can be regarded as a single-pole n-way bilateral switch, and that a CMOS multiplexer can be converted into a demultiplexer by simply transposing the notations of its input and output terminals. An n-way single-pole CMOS bilateral switch is thus often legitimately described as an n-channel multiplexer/demultiplexer. TTL ICs, on the other hand, are based on a unidirectional technology, and a dedicated TTL IC can be designed to act as either a multiplexer or demultiplexer, but (unlike a CMOS bilateral switch) cannot be designed to perform both functions.

Practical ICs

There are three major families of CMOS bilateral switch ICs, and some of these are available in both '4000'-series and '74HC'-series versions. The best known of these families is the '4016/4066' group of Quad bilateral switches; the most popular ICs in this family are the 4016B, 4066B, 74HC4016, and 74HC4066, which all have the functional diagram, etc., shown in *Figure 7.7* and house four independently accessible SPST bilateral switches in a 14-pin DIL package. The 4016B and 74HC4016 use the simple construction shown in *Figure 7.1*, and are recommended for use in sample-and-hold applications where low leakage impedance is of prime

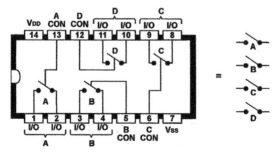

Figure 7.7 Outline and functional diagram common to the 4016B, 4066B, 74HC4016, and 74HC4066 'Quad bilateral switch' ICs, which each act as four independent SPST switches

importance. The 4066B and 74HC4066 use the improved type of construction of *Figure 7.2*, and are recommended for use in all applications where a low ON resistance is of prime importance. *Figure 7.8* lists the major parameter values of these four versions of the quad bilateral switch IC.

PARAMETER	4016B	4066B	74HC4016	74HC4066
Supply voltage (Vs) range [Vᴅᴅ to Vss]	3V to 15V	3V to 15V	2V to 12V	2V to 12V
ON resistance (Vs = 5V) " " (Vs = 10V) " " (Vs = 15V)	520R 250R 200R	250R 120R 80R	70R 30R -	40R 18R -
Bandwidth (Vs = 10V)	54MHz	65MHz	120MHz	120MHz
Crosstalk between any two switches, at 1MHz (Vs = 10V)	-80dB	-50dB	-50dB	-50dB
Max. switch turn ON/OFF delay (Vs = 10V)	20nS	35nS	10nS	10nS

Figure 7.8 Major parameter values of the four main versions of the '4016/4066' family of Quad bilateral switch ICs

One less well-known member of the basic '4016/4066' family of Quad bilateral switch ICs is the 74HC4316, which can be regarded as a variant of the 74HC4016 in which each 'switch' is provided with an integral logic level translator. This IC uses a 16-pin DIL package.

The second major family of ICs is know as the '4051/2/3' group of multichannel multiplexer/ demultiplexer ICs with built-in logic-level conversion. The most popular members of this family are the 4051B and 74HC4051 (see *Figure 7.9*), which are 8-channel multiplexer/demultiplexer ICs, and can each be regarded as a single-pole, 8-way bilateral switch. The ICs have three binary control inputs (A, B, and C) and one INHIBIT input (which disables all switches when biased high). The three binary signals select the one of the 8 channels to be turned on, as shown in the table.

Other popular members of this family are the 4052B and 74HC4052 (see *Figure 7.10*) differential 4-channel multiplexer/demultiplexer ICs, which can each be regarded as a ganged 2-pole 4-way bilateral switch, and the 4053B and 74HC4053 (see *Figure 7.11*) Triple 2-channel multiplexer/demultiplexer ICs, which can each

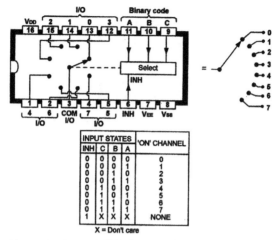

Figure 7.9 Functional diagram and Truth Table common to the 4051B and 74HC4051 8-channel multiplexer/demultiplexer ICs, which act like single-pole 8-way switches

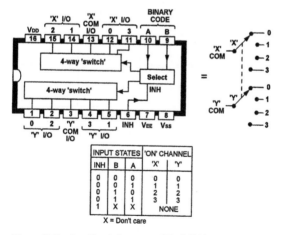

Figure 7.10 Functional diagram and Truth Table common to the 4052B and 74HC4052 differential 4-channel multiplexer/demultiplexer ICs, which act like ganged 2-pole 4-way switches

be regarded as a set of three independent single-pole 2-way bilateral switches. *Figure 7.12* lists the main parameter values of these six popular members of the '4051/2/3' group of ICs.

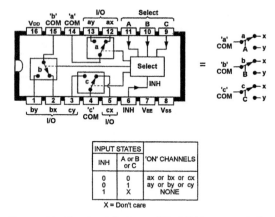

Figure 7.11   Functional diagram and Truth Table common to the 4053B and 74HC4053 Triple 2-channel multiplexer/demultiplexer ICs, which act like three independent single-pole 2-way switches

PARAMETER	5051B	74HC4051	5052B	74HC5052	5053B	74HC5053
Supply voltage (Vs) range (V$_{DD}$ to Vss)	3V to 15V	2V to 12V	3V to 15V	2V to 12V	3V to 15V	2V to 12V
ON resistance (Vs = 5V)	250R	40R	250R	40R	250R	40R
" " (Vs = 10V)	120R'	30R	120R	30R	120R	30R
" " (Vs = 15V)	80R	-	80R	-	80R	-
Bandwidth (Vs = 10V)	20MHz	120MHz	30MHz	120MHz	55MHz	120MHz
Max. turn ON/OFF delay (Vs = 10V)	170nS	18nS	170nS	18nS	140nS	18nS

Figure 7.12   Major parameter values of the six main members of the '4051/2/3' family of multichannel multiplexer/demultiplexer ICs

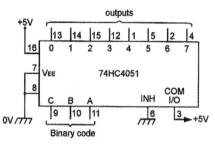

Figure 7.13   Normal connections for using the 74HC4051 as a single-pole digital 8-way switch powered from a 5V supply

The '4051/2/3' family of ICs are quite simple to use. Each IC has three 'power supply' terminals (V_{DD}, V_{SS}, and V_{EE}). In all applications, V_{DD} is taken to the positive supply rail, V_{SS} is grounded, and all digital control signals (for channel select, inhibit, etc.) use these two terminals as their logic reference values, i.e. logic-1 = V_{DD}, and logic-0 = V_{SS}. In digital signal processing applications, terminal V_{EE} is grounded (tied to V_{SS}). In analogue signal processing applications, V_{EE} must be taken to a negative supply rail; ideally, V_{EE} = -(V_{DD}). In all cases, the V_{EE}-to-V_{DD} voltage must be limited to the maximum values shown in *Figure 7.12*, and the INHIBIT terminal is normally tied low. *Figure 7.13* shows an example of these basic design rules

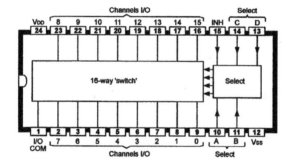

INPUT STATES					'ON' CHANNEL
INH	D	C	B	A	
0	0	0	0	0	0
0	0	0	0	1	1
0	0	0	1	0	2
0	0	0	1	1	3
0	0	1	0	0	4
0	0	1	0	1	5
0	0	1	1	0	6
0	0	1	1	1	7
0	1	0	0	0	8
0	1	0	0	1	9
0	1	0	1	0	10
0	1	0	1	1	11
0	1	1	0	0	12
0	1	1	0	1	13
0	1	1	1	0	14
0	1	1	1	1	15
1	X	X	X	X	NONE

X = Don't care

Figure 7.14 Functional diagram and Truth Table of the 4067B 16-channel multiplexer/demultiplexer IC, which acts like a single-pole 16-way switch

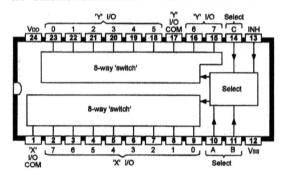

Figure 7.15 Functional diagram and Truth Table of the 4097B differential 8-channel multiplexer/demultiplexer IC, which acts like a ganged 2-pole 8-way switch

applied to a 74HC4051 IC, which is shown powered from a 5V single-ended supply and wired as a simple digital single-pole 8-way switch, with the switch position selected via the C–B–A terminals.

Less well-known members of the basic '4051/2/3' family of ICs are the 74HC4351, 74HC4352, and 74HC4353, which can be regarded as a variant of the 74HC4051, 74HC4052, and 74HC4053 respectively, but each incorporate a built-in ADDRESS latch.

The final family of devices consists of the 4067B and 4097B multiplexer/demultiplexer types (*Figures 7.14 and 7.15*). These devices can be used in both analogue and digital applications, but do not feature built-in logic-level conversion. The 4067B is a 16-channel device, and can be regarded as a single-pole 16-way bilateral switch. The 4097B is a differential 8-channel device, and can be regarded as a ganged 2-pole 8-way bilateral switch. Each IC is housed in a 24-pin dual-in-line package.

Using '4016/4066' ICs

The '4016/4066' family of ICs are the most popular types of CMOS bilateral switches, and the rest of this chapter is devoted to showing various ways of using ICs from this particular family. Note that the following notes, although specifically dedicated to the 4016B or 4066B ICs, are equally applicable to the 74HC4016 and 74HC4066 types.

The 4016B and 4066B ICs are very versatile ICs, but a few simple precautions must be taken when using them, as follows:

1. Input and switching signals must never be allowed to rise above V_{DD} or fall below V_{SS}.
2. Each unused section of the IC must be disabled (see *Figure 7.16*) either by taking its control terminal to V_{DD} and wiring one of its switch terminals to V_{DD} or V_{SS}, or by taking all three terminals to V_{SS}.

Figures 7.17 to *7.22* show some simple applications of the 4066B (or 4016B). *Figure 7.17* shows the device used to implement the four basic switching functions of SPST, SPDT, DPST and DPDT. *Figure 7.17(a)* shows the SPST connections, which have already been described. The SPDT function (*Figure 7.17(b)*) is implemented by wiring a CMOS inverter stage between the IC1a and IC1b control terminals. The DPST switch (*Figure 7.17(c)*) is simply two SPST switches sharing a common control terminal, and the DPDT switch (*Figure*

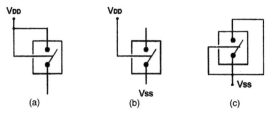

Figure 7.16 Unused sections of the 4066B must be disabled, using any one of the connections shown here

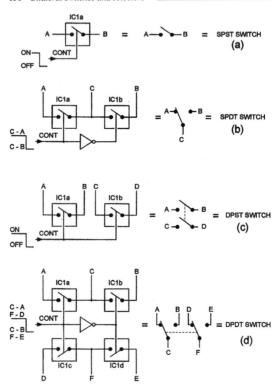

Figure 7.17 Implementation of the four basic switching functions via the 4066B (IC1)

7.17(d)) is two SPDT switches sharing an inverter stage in the control line.

Note that the switching functions of *Figure 7.17* can be expanded or combined in any desired way by using more IC stages. Thus, a 10-pole 2-way switch can be made by using five of the *Figure 7.17*(d) circuits with their control lines tied together.

Each 4066B bilateral switch has a typical ON resistance of about 90R (when powered from a 12V supply). *Figure 7.18* shows how four standard switch elements can be wired in parallel to make a single switch with a typical ON resistance of only 22.5 ohms.

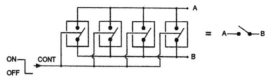

Figure 7.18 This SPST switch has a typical ON resistance of only 22.5 ohms

Figures 7.19 to *7.22* show ways of using a bilateral switch as a self-latching device. In these circuits the switch current flows to ground via R3, and the control terminal is tied to the top of R3 via R2. Thus, in *Figure 7.19*, when PB1 is briefly closed the control terminal is pulled to the positive rail and the bilateral switch closes. With the bilateral switch closed, the top of R3 is at supply-line potential and, since the control terminal is tied to R3 via R2, the bilateral switch is thus latched on. Once latched, the switch can only be turned off again by briefly closing PB2, at which point the bilateral switch opens and the R3 voltage falls to zero. Note here that LED1 merely indicates the state of the bilateral switch, and R1 prevents supply-line shorts if PB1 and PB2 are both closed at the same time.

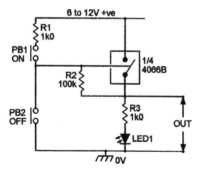

Figure 7.19 Latching push-button switch

Figure 7.20 shows how the above circuit can be made to operate as a latching touch-operated switch by increasing R2 to 10M and using R4–C1 as a 'hum' filter.

Figures 7.21 and *7.22* show alternative ways of using the *Figure 7.19* circuit to connect power to external

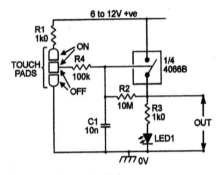

Figure 7.20   Latching touch switch

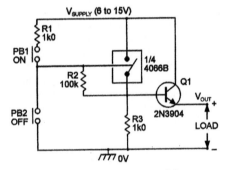

Figure 7.21   Latching push-button power switch

circuitry. The *Figure 7.21* circuit connects the power via a voltage follower stage, and the *Figure 7.22* design connects the power via a common-emitter amplifier.

Digital Control Circuits

Bilateral switches can be used to digitally control or vary effective values of resistance, capacitance, impedance, amplifier gain, oscillator frequency, etc., in any desired number of discrete steps. *Figure 7.23* shows how the four switches of a single 4066B can, by either shorting or not shorting individual resistors in a chain,

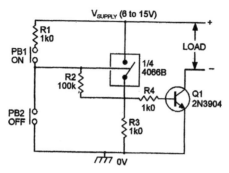

Figure 7.22   Alternative version of the latching power switch

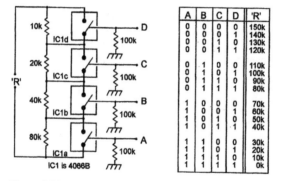

A	B	C	D	'R'
0	0	0	0	150k
0	0	0	1	140k
0	0	1	0	130k
0	0	1	1	120k
0	1	0	0	110k
0	1	0	1	100k
0	1	1	0	90k
0	1	1	1	80k
1	0	0	0	70k
1	0	0	1	60k
1	0	1	0	50k
1	0	1	1	40k
1	1	0	0	30k
1	1	0	1	20k
1	1	1	0	10k
1	1	1	1	0k

Figure 7.23   16-step digital control of resistance. 'R' can be varied from zero to 150k in 10k steps

be used to vary the effective total value of the resistance chain in 16 digitally controlled steps of 10k each. In practice, the step magnitudes of this basic circuit can be given any desired value (determined by the value of the smallest resistor) so long as the four resistors are kept in the ratio 1–2–4–8. The number of 'steps' can be increased by adding more resistor/switch stages; thus, a 6-stage circuit (with resistors in the ratio 1–2–4–8–16–32) will give resistance variation in 64 steps.

Figure 7.24 shows how four switches can be used to make a digitally-controlled capacitor that can be varied in 16 steps of 1n0 each. Again, the circuit can be expanded to give more 'steps' by simply adding more stages.

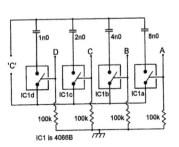

A	B	C	D	'C'
0	0	0	0	0
0	0	0	1	1n0
0	0	1	0	2n0
0	0	1	1	3n0
0	1	0	0	4n0
0	1	0	1	5n0
0	1	1	0	6n0
0	1	1	1	7n0
1	0	0	0	8n0
1	0	0	1	9n0
1	0	1	0	10n
1	0	1	1	11n
1	1	0	0	12n
1	1	0	1	13n
1	1	1	0	14n
1	1	1	1	15n

Figure 7.24   16-step digital control of capacitance. 'C' can be varied from zero to 15n in 1n0 steps

Note in the *Figure 7.23* and *7.24* circuits that the resistor/capacitor values can be controlled by operating the 4066B switches manually, or automatically via simple logic networks, or via up/down counters (see Chapter 12), or via microprocessor control, etc.

The circuits of *Figures 7.23* and *7.24* can be combined in a variety of ways to make digitally controlled impedance and filter networks, etc. *Figure 7.25*, for example, shows two alternative ways of using them to make a digitally controlled 1st-order low-pass filter.

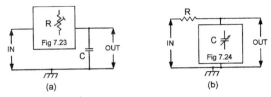

Figure 7.25   Alternative ways of using *Figure 7.23* or *7.24* to make a digitally-controlled 1st-order low-pass filter

Digital control of amplifier gain can be obtained by hooking the *Figure 7.23* circuit into the feedback or input path of a standard op-amp inverter circuit, as shown in *Figures 7.26* and *7.27*. The gain of such a circuit equals R_f/R_{in}, where R_f is the feedback resistance and R_{in} is the input resistance. Thus, in *Figure 7.26*, the gain can be varied from zero to unity in 16 steps of '1/15th' each, giving a sequence of 0/15 (i.e. zero), 1/15, 2/15, etc., up to 14/15 and (finally) 15/15 (i.e. unity).

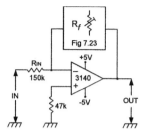

Figure 7.26 Digital control of gain, using the *Figure 7.23* circuit. Gain is variable from zero to unity in 16 steps

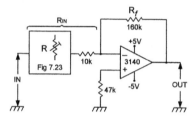

Figure 7.27 Digital control of gain, using the *Figure 7.23* circuit. Gain is variable from unity to ×16, in 16 steps

In the *Figure 7.27* circuit the gain can be varied from unity to ×16 in 16 steps, giving a gain sequence of 1, 2, 3, 4, 5, etc. Note in both of these circuits that the op-amp uses a split power supply, so the 4066B control voltage must switch between the negative and positive supply rails.

Figure 7.28 shows how the *Figure 7.23* circuit can be used to vary the frequency of a 555 astable oscillator in 16 discrete steps. Finally, *Figure 7.29* shows how three

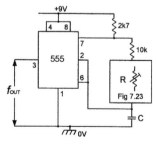

Figure 7.28 Digital control of 555 astable frequency, in 16 steps

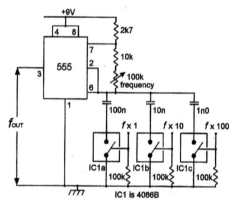

Figure 7.29   Digital control of decade range selection of a 555 astable

bilateral switches can be used to implement digital control of decade range selection of a 555 astable oscillator. Here, only one of the switches must be turned on at a time. Naturally, the circuits of *Figures 7.28* and *7.29* can easily be combined, to form a wide-range oscillator that can be digitally controlled via a microprocessor or other device.

Synthesized Multi-gang Pots, etc.

One very useful application of the bilateral switch is in synthesizing multi-gang rheostats, potentiometers (pots), and variable capacitors in a.c. signal processing circuitry. The synthesizing principle is quite simple and is illustrated in *Figure 7.30*, which shows the circuit of a 4-gang 10k-to-100k rheostat for use at signal frequencies up to about 15kHz.

Here, the 555 is used to generate a 50kHz rectangular waveform that has its mark–space (M–S) ratio variable from 11:1 to 1:11 via RV1, and this waveform is used to control the switching of the 4066B stages. All of the 4066B switches are fed with the same control waveform, and each switch is wired in series with a

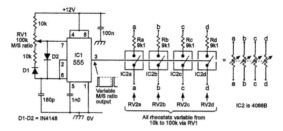

Figure 7.30 Synthesized precision 4-gang 'rheostat'

range resistor (Ra, Rb, etc.), to form one gang of the 'rheostat' between the 'aa', 'bb', etc., terminals.

Remembering that the switching rate of this circuit is fast (50kHz) relative to the intended top signal frequency (15kHz), it can be seen that the *mean* or effective value (when integrated over a few switching cycles) of each 'rheostat' resistance can be varied via M–S ratio control RV1.

Thus, if IC2a is closed for 90% and open for 10% of each duty cycle (M–S ratio = 9:1), the apparent (mean) value of the 'aa' resistance will be 10% greater than Ra, i.e. 10k. If the duty cycle is reduced to 50%, the apparent Ra value will double, to 18.2k. If the duty cycle is further decreased, so that IC2a is closed for only 10% of each duty cycle (M–S ratio = 1:9), the apparent value of Ra will increase by a decade, to 91k. Thus, the apparent value of each gang of the 'rheostat' can be varied via RV1.

There are some important points to note about this circuit. First, it can be given any desired number of gangs by simply adding an appropriate number of switch stages and range resistors. Since all switches are controlled by the same M–S ratio waveform, perfect tracking is assured between the gangs. Individual gangs can be given different ranges, without affecting the tracking, by giving them different range resistor values. Also, note that the 'sweep' range and 'law' of the rheostats can be changed by simply altering the characteristics of the M–S ratio generator, but that the switching control frequency *must* be far higher than the top signal frequency that is to be handled.

The 'rheostat' circuit of *Figure 7.30* can be made to function as a multi-gang variable capacitor by using ranging capacitors in place of ranging resistors. In this case, however, the apparent capacitance value decreases as the duty cycle is decreased.

The *Figure 7.30* principle can be expanded, to make synthesized multi-gang pots, by using the basic technique shown in *Figure 7.31*. Here, in each gang, two 'rheostats' are wired in series but have their switch control signals fed in anti-phase, so that one rheostat value increases as the other decreases, thus giving a variable potential-divider action. This basic circuit can be expanded, to incorporate any desired number of gangs, by simply adding more double-rheostat stages.

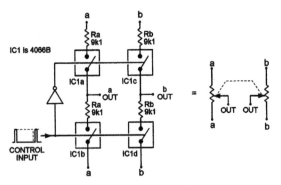

Figure 7.31   Synthesized precision 2-gang 'pot'

Miscellaneous Applications

To complete this look at the CMOS bilateral switch, *Figures 7.32* and *7.33* show a couple of miscellaneous fast-switching applications of the device. *Figure 7.32* shows how the bilateral switch can be used as a sample-and-hold element. The 3140 CMOS op-amp is used as a voltage follower and has a near-infinite input impedance, and the 4016B switch also has a near-infinite impedance when open. Thus, when the 4016B is

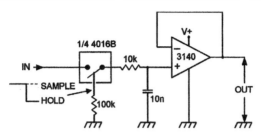

Figure 7.32 Using a bilateral switch as a sample-and-hold element

closed, the 10n capacitor rapidly follows all variations in input voltage, but when the switch opens the prevailing capacitance charge is 'stored' and the resulting voltage remains available at the op-amp output.

Finally, *Figure 7.33* shows a bilateral switch used in a linear ramp generator. Here, the op-amp is used as an integrator, with its non-inverting pin biased at 5V via R1–R2, so that a constant current of 5μA flows into the inverting pin via R. When the bilateral switch is open, this current linearly charges capacitor C, causing a rising ramp to appear at the op-amp output. When the bilateral switch closes, C is rapidly discharged via R3 and the output switches down to 5V.

The basic *Figure 7.33* circuit is quite versatile. The switching can be activated automatically via a free-running oscillator or via a voltage-trigger circuit. Bias levels can be shifted by changing the R1–R2 values or can be switched automatically via another 4066B stage, enabling a variety of ramp waveforms to be generated.

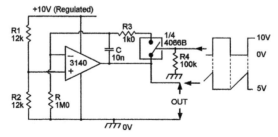

Figure 7.33 Using the 4066B to implement a ramp generator circuit

8 Pulse Generator Circuits

Digital ICs must, in general, be driven or clocked by
clean input waveforms that switch between normal logic
levels and have very sharp leading and trailing edges;
such waveforms can be generated by suitably using
standard logic ICs, or via dedicated waveform-generator
ICs. This chapter explains digital waveform generator
basics, and shows a variety of practical pulse waveform
'clock' generator circuits; Chapter 9 continues the theme
by taking a detailed look at practical squarewave
generator circuits.

Waveform Generator Basics

Most digital waveform generator circuits fit into one or
other of the four basic categories shown in *Figure 8.1(a)*
to *(d)*. Schmitt trigger circuits *(a)* produce an output that
switches abruptly between logic-0 and logic-1 values
whenever an input signal goes above or below preset
instantaneous voltage levels. Circuits of this type can
thus be used in waveform-shaper applications, such as
converting a sine or ramp input waveform into a square
or pulse-shaped output, etc.

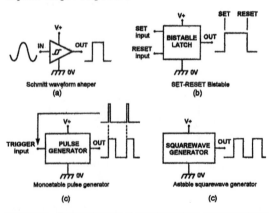

Figure 8.1   Four basic types of digital waveform generator circuit

Simple bistable circuits *(b)* have two input terminals, known as SET and RESET. The circuit's output can be latched into the logic-1 state by applying a suitable command signal (usually a brief logic-1 pulse) to the SET input terminal; the output can then be latched into the alternative logic-0 state only by applying a suitable command signal to the RESET terminal, and so on. 'Latch' circuits of this type are useful in a variety of digital signal-conditioning applications.

Monostable circuits *(c)* have an output that switches high on the arrival of an input trigger signal, but then automatically switches back to the low state again after some preset time delay. Circuits of this basic type thus act as triggered pulse generators.

Astable circuits *(d)* have an output that switches into the logic-1 state for a preset period, then switches into the logic-0 state for a second preset period, and then switches back into the first state again, and so on. Circuits of this basic type thus generate a free-running squarewave output.

In practice, all of these basic circuits can be elaborated into more complex forms. The simple monostable, for example, can be elaborated into a 'resettable' form, in which the output pulse can be terminated prematurely via a 'reset' signal, or into a 'retriggerable' form, in which a new monostable timing period is initiated each time a new trigger pulse arrives. Astable circuits may be completely free-running, or may be gated types which operate only in the presence of a suitable gate signal, etc.

Schmitt Edge Detector Circuits

A number of Schmitt waveform shaper, switch 'debouncer', and switch-on pulse-generator circuits, etc., have already been described in this book (see, for example, *Figures 3.4*, *3.6*, *4.19*, and *6.29* to *6.34*). Another useful application of the Schmitt element is as an 'edge'-detector that produces a sharp output pulse on the arrival of either the leading or the trailing edge of a digital input waveform. In most practical applications the precise width of the output pulse is non-critical.

The basic way of making an edge detector is to feed the digital input waveform to a Schmitt element via a short-time-constant C–R differentiation network of the type shown on *Figure 8.2*, in which – when fed with a 3V squarewave input signal – the network's output switches abruptly to +3V on the arrival of the squarewave's rising edge but then decays rapidly to zero volts as C1 charges up via R1, and switches abruptly to –3V on the arrival of the squarewave's falling edge but then decays rapidly to zero volts as C1 discharges via R1, and so on. In a practical Schmitt edge detector circuit, these output 'spike' waveforms must be fed to the Schmitt stage in such a way that only one of the two spikes has any practical effect. The Schmitt element may be of either the inverting or the non-inverting type, depending on the required polarity of the output pulse waveform; in either case, the resulting circuit is known as a half-monostable or 'half-mono' pulse generator.

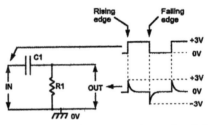

Figure 8.2   A simple C–R coupler can be used to distinguish between the 'rising' and 'falling' edges of a digital waveform

Figure 8.3(a) shows how the C–R coupler can be used in conjunction with a TTL Schmitt inverter to make a pulse-generating rising-edge detector. Here, R1 normally ties the Schmitt input low, so that its output is at logic-1, but on the arrival of the input rising edge the Rl positive spike drives the Schmitt output briefly to logic-0, so that it generates a negative-going output pulse. Note that, since the Schmitt's input is normally tied low, the negative 'falling edge' input spikes have no effect on the circuit. *Figure 8.3(b)* shows how the circuit can be made

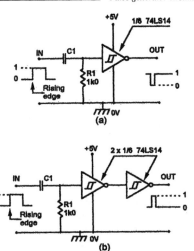

Figure 8.3 TTL rising-edge detectors giving *(a)* negative-going or *(b)* positive-going output pulses

to give a positive-going output pulse by simply taking the output via another Schmitt inverter stage.

Figure 8.4(a) shows the basic TTL Schmitt circuit modified so that it generates a positive-going output

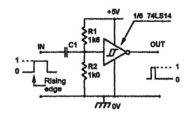

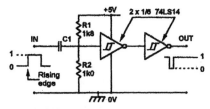

Figure 8.4 TTL falling-edge detectors giving *(a)* positive-going or *(b)* negative-going output pulses

pulse on the arrival of a falling (rather than rising) input edge. Here, the Schmitt's input is normally biased above its 1.6V upper threshold value via the R1–R2 divider, so it ignores the effects of positive 'rising edge' spikes, and its output is normally at logic-0. But on the arrival of each negative 'falling edge' spike its input is driven below the Schmitt's 0.8V lower threshold value, and it generates a positive-going output pulse. *Figure 8.4(b)* shows the circuit modified to give a negative-going output pulse via an additional Schmitt inverter stage. (In practice, the output pulse widths of the *Figure 8.3* and *8.4* circuits vary greatly with individual ICs, but approximate 1μs per nF of C1 value).

CMOS Schmitt elements are simpler to use than TTL types in basic edge-detector applications. This is because the built-in protection diodes that are connected to each input terminal can be used to perform the basic 'spike discriminator' action decribed earlier. *Figure 8.5* shows two ways of making a CMOS rising edge-detecting half-mono circuit. Here, the input of the Schmitt is tied to ground via resistor R, and C–R have a time constant that is short relative to the period of the input waveform. The leading edge of the input signal is thus converted into the spike waveform shown, and this spike is then converted into a good clean pulse waveform via the Schmitt element. The circuit generates a positive-going output pulse if a non-inverting Schmitt is used (*Figure 8.5(a)*), or a negative-going pulse if an inverting Schmitt is used (*Figure 8.5(b)*); in either case, the output pulse has a period (P) of roughly 0.7CR.

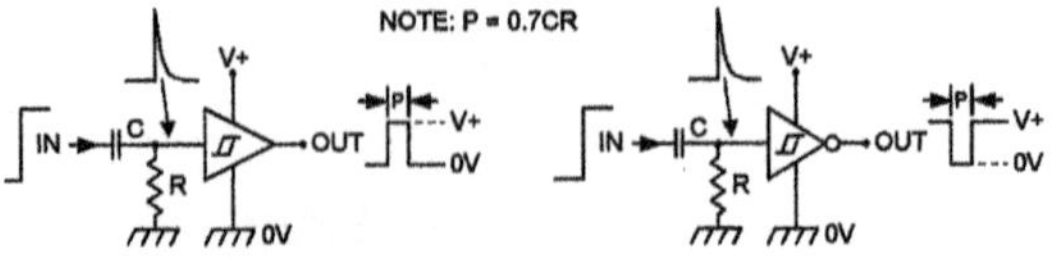

Figure 8.5 CMOS rising-edge detector circuits giving *(a)* positive and *(b)* negative output pulses

Figure 8.6 shows a falling edge detecting version of the half-mono. In this case the Schmitt input is tied to the positive supply rail via R, and C–R again has a short time constant. The circuit generates a positive-going

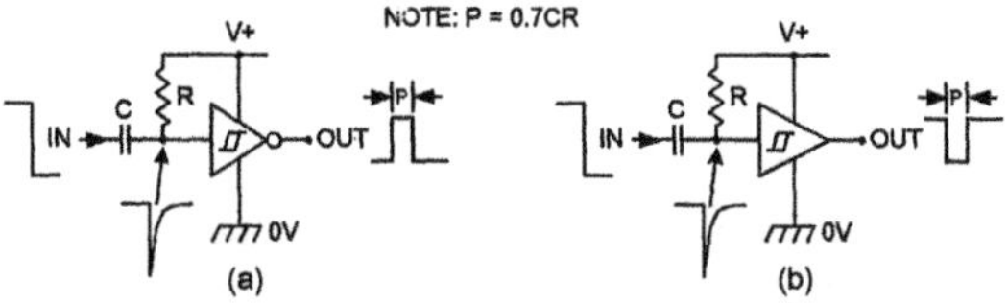

Figure 8.6 CMOS falling-edge detector circuits giving *(a)* positive and *(b)* negative output pulses

output pulse if an inverting Schmitt is used (*Figure 8.6(a)*), or a negative-going pulse if a non-inverting Schmitt is used. The output pulse has a period of roughly 0.7CR.

Before leaving this subject, note from *Figure 8.2* that a positive-going input pulse has a rising leading edge and a falling trailing edge, so in this case a rising-edge detector acts as a 'leading-edge' detector, but that a negative-going input pulse has a falling leading edge and a rising trailing edge, so in this case a rising-edge detector acts as a 'trailing-edge' detector.

Bistable Waveform Generators

Strictly speaking, a clock generator can be any circuit that generates a clean (noiseless, with sharp leading and trailing edges) waveform suitable for clocking fast digital counter/divider circuitry, etc. One particularly useful cicuit of this type is the simple R–S (Reset–Set) bistable (also known as the R–S flip-flop), which can be used to deliver a single clock pulse each time the available one of its two input terminals is activated, either electronically or via a push-button switch, etc.

The bistable waveform generator is a circuit that can have its output set to either the logic-1 or logic-0 state by applying a suitable control signal to either its SET or RESET input terminal. The simplest way to make an S–R (or R–S) circuit of this type is to wire a pair of normal or Schmitt inverter elements into a feedback loop as shown in the TTL example of *Figure 8.7*. If RESET switch S1 is briefly closed it shorts the A input

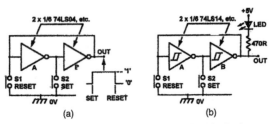

Figure 8.7 These manually-triggered TTL bistable circuits can be built from either *(a)* normal or *(b)* Schmitt inverter elements

low, driving the A output and B input high, thus making the B output go low and lock the A input into the low state irrespective of the subsequent state of S1. The circuit thus latches into this RESET state, with its output at logic-0, until SET switch S2 is briefly closed, at which point the B output and A input are driven high, thus making the A output go low and lock the B input into the low state irrespective of the subsequent state of S2. The circuit thus latches into this SET state, with its output at logic-1, until the S1 RESET switch is next closed, at which point the whole sequence starts to repeat again.

Note that each time S1 (or S2) is operated it places a short between the B (or A) output and ground, but that the resultant output current is internally limited to safe values by the TTL inverter's totem-pole output stage and only effectively flows for the few nanoseconds that are taken (by the circuit) to switch the inverter into its latched 'output low' state. These apparently brutal circuits are thus, in reality, soundly engineered and are delightfully cheap and effective designs that generate perfectly 'bounce-free' output switching waveforms. If desired, the circuit's output state can be monitored by an LED connected as shown in *Figure 8.7(b)*, so that the LED glows when the output is in the 'RESET' (low) state. The basic design can be modified for operation via a single toggle switch by connecting it as shown in the circuit of *Figure 8.8*, which generates a perfectly reliable and bounce-free 'toggle' output waveform.

The basic *Figure 8.7* and *8.8* circuits are not really suitable for activation by electronic trigger-pulse

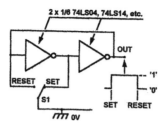

Figure 8.8 This TTL 'bistable toggle switch' gives perfect waveform generation from a normal toggle switch

signals, etc., but can be adapted for this operation by replacing the simple inverters with 2-input NAND or NOR gates connected as shown in *Figure 8.9*, so that the control inputs are not subject to odd loading effects. If NAND gates are used, both inputs must normally be biased high, and the output is SET or RESET by briefly pulling the appropriate control input low. If NOR gates are used, both inputs must normally be biased low, and the output is SET or RESET by briefly driving the appropriate input high.

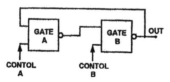

Figure 8.9 A bistable latch can be built by interconnecting two NAND or NOR 2-input gates as shown

Figure 8.10 shows ways of using a TTL NAND-type SET–RESET (S–R) bistable as a manually triggered 'toggle' waveform generator that is activated via either *(a)* two push-button switches or *(b)* a single toggle switch. These diagrams act as handy 'wiring' guides, but note that the conventional way of drawing the NAND bistable is as in *Figure 8.11(a)*. The NAND bistable is a standard logic element, and is usually called an S–R (or R–S) bistable or flip-flop; *(b)* shows its international circuit symbol, in which the little negation circle on each input indicates that it is activated by logic-0 levels, and which has both normal (Q) and inverted (NOT–Q) outputs; *(c)* shows the element's Truth Table, which

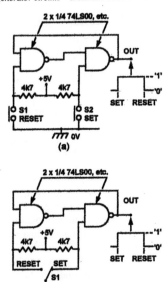

Figure 8.10   TTL NAND bistables used as manually-triggered 'toggle' waveform generators

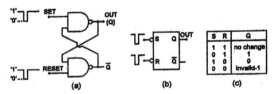

Figure 8.11   Conventional schematic *(a)*, symbol *(b)*, and Truth Table of the NAND-type S–R bistable or flip-flop

indicates that both inputs must normally be biased high (to logic-1), and that the element goes into an invalid state – with both outputs set to logic-1 – if both inputs are low at the same time. *Figures 8.12* and *8.13* show practical CMOS versions of pulse-triggered and manually triggered bistables, together with their output waveforms, drawn in the 'conventional' manner.

Figure 8.14 shows ways of using a TTL NOR-type S–R bistable as a manually triggered 'toggle' waveform

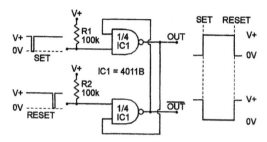

Figure 8.12 CMOS pulse-triggered NAND bistable

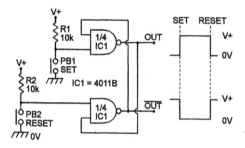

Figure 8.13 CMOS manually-triggered NAND bistable

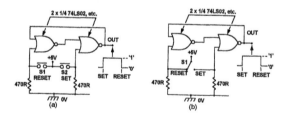

Figure 8.14 TTL NOR bistables used as manually triggered 'toggle' waveform generators

generator that is activated via either *(a)* two push-button switches or *(b)* a single toggle switch. *Figure 8.15(a)* shows the conventional way of drawing the NOR bistable circuit, which is also a standard logic element; *(b)* shows the element's international circuit symbol, and *(c)* shows its Truth Table, which indicates that both inputs must normally be biased low (to logic-0), and that the element goes into an invalid state – with both outputs set to logic-0 – if both inputs are high at the

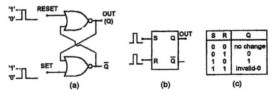

Figure 8.15 Conventional schematic *(a)*, symbol *(b)*, and Truth Table *(c)* of the NOR-type S–R (or R–S) bistable or flip-flop

same time. *Figures 8.16* and *8.17* show practical CMOS versions of pulse-triggered and manually triggered bistables, together with their output waveforms, drawn in the 'conventional' manner.

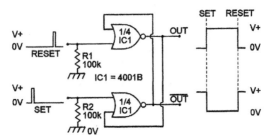

Figure 8.16 CMOS pulse-triggered NOR bistable

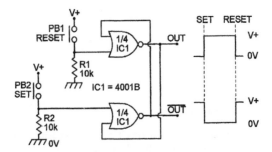

Figure 8.17 CMOS Manually-triggered NOR bistable

If more than two TTL S–R bistables are needed in any application, the most cost-effective way to get them is from a 74LS279 Quad S–R latch IC, which is shown in *Figure 8.18*, together with its Truth Table. This IC

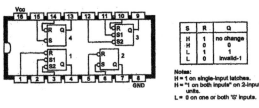

S	R	Q
H	1	no change
H	0	0
L	1	1
L	0	Invalid-1

Notes:
H = 1 on single-input latches.
H = "1 on both inputs" on 2-input units.
L = 0 on one or both 'S' inputs.

Figure 8.18 Functional diagram of the 74LS279 Quad S–R latch IC

houses two normal NAND-type bistables (elements 2 and 4), plus two NAND-type bistables which each have two SET (S1 and S2) inputs. *Figure 8.19* shows basic ways of using 74LS279 elements as manually triggered SET–RESET latches; *(a)* shows how to use a simple '2' or '4' element in this mode; *(b)* and *(c)* show how to use '2-SET'-input elements by either tying both SET terminals together, or by tying one input to logic-1 and applying the SET signal to the other.

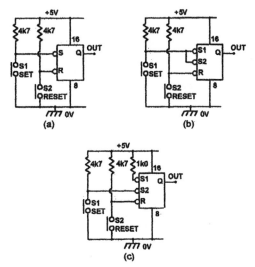

Figure 8.19 Basic ways of using 74LS279 elements as SET–RESET latches

Monostable Pulse Generator Basics

A monostable ('mono') or 'one-shot' pulse generator is a circuit that generates a single high-quality output pulse of some specific width or period p on the arrival of a suitable trigger signal. In a standard monostable circuit the arrival of the trigger signal initiates an internal timing cycle which causes the monostable output to change state at the start of the timing cycle, but to revert back to its original state on completion of the cycle, as shown in *Figure 8.20*. Note that once a timing cycle has been initiated the standard monostable circuit is immune

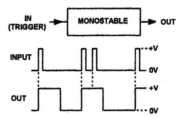

Figure 8.20 A standard monostable generates an accurate output pulse on the arrival of a suitable trigger signal

to the effects of subsequent trigger signals until its timing period ends naturally, but it then needs a certain 'recovery' time (usually equal to p or greater) to fully reset before it can again generate an *accurate* triggered output pulse; it thus cannot normally generate accurate pulse output waveforms with duty cycles greater than about 50%. This type of circuit is sometimes modified by adding a RESET control terminal, as shown in *Figure 8.21*, to enable the output pulse to be terminated or aborted at any time via a suitable command signal.

Another variation of the monostable is the 'retriggerable' circuit. Here, the trigger signal actually resets the mono and almost simultaneously initiates a new pulse-generating timing cycle, as shown in *Figure 8.22*, so that each new trigger signal initiates a new timing cycle, even if the trigger signal arrives in the midst of an existing cycle. This type of circuit has a very

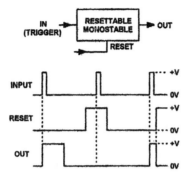

Figure 8.21 The output pulse of a resettable mono can be aborted by a suitable reset pulse

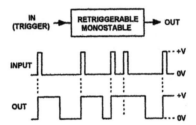

Figure 8.22 A retriggerable mono starts a new timing cycle on the arrival of each new trigger signal

short recovery time, and can generate accurate pulse output waveforms with duty cycles up to almost 100%.

Most monostables are 'edge' triggered, i.e. their pulse generation cycle is initiated (fired) by the arrival of the trigger signal's rising or falling edge; this type of mono needs a well-shaped trigger signal, with fast edges. Some monostables, however, are 'voltage level' triggered via a Schmitt input stage, and fire when the voltage reaches a predetermined value; this type of mono can be fired by any shape of input signal.

Thus, the circuit designer may use an edge-triggered or level-triggered standard mono, resettable mono, or retriggerable mono to generate triggered output pulses, the 'type' decision depending on the specific circuit design requirements. In many cases, if only modest pulse-width precision is required, standard monostable circuits can be built using low-cost CMOS logic or flip-flop ICs. If greater precision or sophistication is required, the circuits can be built using one of the many TTL or CMOS dedicated pulse-generator ICs that are available.

Simple CMOS Monostable Circuits

The cheapest way of making a standard or a resettable monostable pulse generator is to use a 4001B Quad 2-input NOR gate or a 4011B Quad 2-input NAND gate IC (or a '74HC'-series equivalent) in one of the configurations shown in *Figures 8.23* to *8.26*. Note, however, that the output pulse widths of these circuits are subject to fairly large variations between individual ICs and with variations in supply rail voltage, and these circuits are thus not suitable for use in high-precision applications.

Figures 8.23 and *8.24* show alternative versions of the standard monostable circuit, each using only two of the four available gates in the specified CMOS package. In these circuits the duration of the output pulse is determined by the C1–R1 values, and approximates 0.7 × C1 × R1. Thus, when R1 has a value of 1M5 the pulse period approximates one second per µF of C1 value. In practice, C1 can have any value from about 100pF to a few thousand µF, and R1 can have any value from 4k7 to 10M.

An outstanding feature of these circuits is that the input trigger signal can be direct coupled and its duration has little effect on the length of the generated output pulse. The NOR version of the circuit (*Figure 8.23*) has a

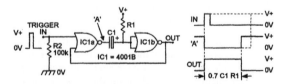

Figure 8.23 2-gate NOR monostable is triggered by a positive-going signal and generates a positive-going output pulse

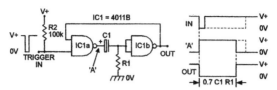

Figure 8.24 2-gate NAND monostable is triggered by a negative-going signal and generates a negative-going output pulse

normally low output and is triggered by the edge of a positive-going input signal, and the NAND version (*Figure 8.24*) has a normally high output and is triggered by the edge of a negative-going input signal. Another feature is that the pulse signal appearing at point 'A' has a period equal to that of either the output pulse or the input trigger pulse, whichever is the greater of the two. This feature is of value when making pulse-length comparators and over-speed alarms, etc.

The operating principle of these monostable circuits is fairly simple. Look first at the case of the *Figure 8.23* circuit, in which IC1a is wired as a NOR gate and IC1b is wired as an inverter. When this circuit is in the quiescent state the trigger input terminal is held low by R2, and the output of IC1b is also low. Thus, both inputs of IC1a are low, so IC1a output is forced high and C1 is discharged.

When a positive trigger signal is applied to the circuit the output of IC1a is immediately forced low and (since C1 is discharged at this moment) pulls IC1b input low and thus drives IC1b output high; IC1b output is coupled back to the IC1a input, however, and thus forces IC1a output to remain low irrespective of the prevailing state of the trigger signal. As soon as IC1a output switches low, C1 starts to charge up via R1 and, after a delay determined by the C1–R1 values, the C1 voltage rises to such a level that the output of IC1b starts to swing low, terminating the output pulse. If the trigger signal is still high at this moment, the pulse terminates non-regeneratively, but if the trigger signal is low (absent) at this moment the pulse terminates regeneratively.

The *Figure 8.24* circuit operates in a manner similar to that described above, except that IC1a is wired as a NAND gate, with its trigger input terminal tied to the positive supply rail via R2, and the R1 timing resistor is taken to ground.

In the *Figure 8.23* and *8.24* circuits the output is direct-coupled back to one input of IC1a to effectively maintain a 'trigger' input once the true trigger signal is removed, thereby giving a semi-latching circuit operation. These circuits can be modified so that they act as resettable monostables by simply providing them with a means of breaking this feedback path, as shown in *Figures 8.25* and *8.26*. Here, the feedback connection from IC1b output to IC1a input is made via R3. Consequently, once the circuit has been triggered *and the original trigger signal has been removed* each circuit can be reset by forcing the feedback input of IC1a to its normal quiescent state via push-button switch PB1. In practice, PB1 can easily be replaced by a transistor or CMOS switch, etc., enabling the RESET function to be accomplished via a suitable reset pulse.

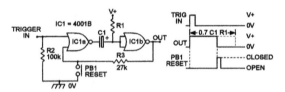

Figure 8.25   Resettable NOR-type monostable

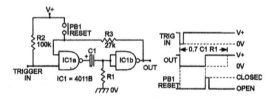

Figure 8.26   Resettable NAND-type monostable

CMOS 'flip-flop' Monostable Circuits

Medium-accuracy monostables can also be built by using standard edge-triggered CMOS flip-flop ICs such as the 4013B dual D-type or the 4027B dual JK-type (both of these ICs are described in detail in Chapter 10) in the configurations shown in *Figures 8.27* and *8.28*. Both of these circuits operate in the same basic way, with the IC wired in the frequency divider mode by suitable connection of its control terminals (DATA and SET in the 4013B, and J, K, and SET in the 4027B), but with the Q terminal connected back to RESET via a C–R time-delay network. The operating sequence of each circuit is as follows.

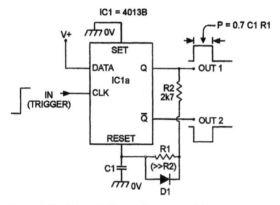

Figure 8.27 D-type flip-flop used as a monostable

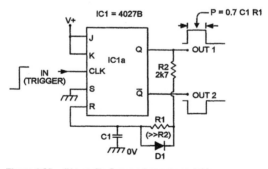

Figure 8.28 JK-type flip-flop used as a monostable

When the circuit is in its quiescent state its Q output pin is low and discharges timing capacitor C1 via R2 and the parallel combination D1–R1. On the arrival of a sharply rising leading edge on the clock terminal the Q output flips high, and C1 starts to charge up via the series combination R1–R2 until eventually, after a delay determined mainly by the C1–R1 values (R1 is large relative to R2), the C1 voltage rises to such a value that the flip-flop is forced to reset, driving the Q terminal low again. C1 then discharges rapidly via R2 and D1–R1, and the circuit is then ready to generate another pulse on the arrival of the next trigger signal.

The timing period of the *Figure 8.27* and *8.28* circuits approximates $0.7 \times C1 \times R1$, and the 'reset' period (the time taken for C1 to discharge at the end of each pulse) approximates $C1 \times R2$. In practice, R2 is used mainly to prevent degradation of the trailing edge of the pulse waveform as C1 discharges; R2 can be reduced to zero if this degradation is acceptable. Note that the circuit generates a positive-going output pulse at Q, and a negative-going pulse (which is not influenced by the R2 value) at NOT-Q.

The *Figure 8.27* and *8.28* circuits can be made resetable by connecting C1 to the RESET terminal via one input of an OR gate and using the other input of the OR gate to accept the external RESET signal. *Figure 8.29* shows how the 4027B circuit can be so modified.

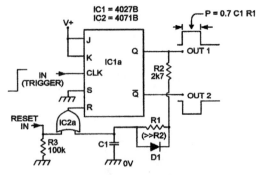

Figure 8.29   Resettable JK-type monostable

Finally, *Figure 8.30* shows how the 4027B can be used to make a retriggerable monostable in which the pulse period restarts each time a new trigger pulse arrives. Note that the input of this circuit is normally high, and that the circuit is actually triggered on the trailing

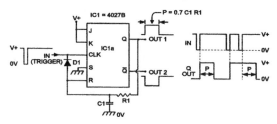

Figure 8.30 Retriggerable JK-type monostable

(rising) edge of a negative-going input pulse. The circuit operates as follows.

At the start of each timing cycle the input trigger pulse switches low and rapidly discharges C1 via D1 and then, a short time later, the trigger pulse switches high again, releasing C1 and simultaneously flipping the Q output high. The timing cycle then starts in the normal way, with C1 charging via R1 until the C1 voltage rises to such a level that the flip-flop resets, driving the Q output low again and slowly discharging C1 via R1. If a new trigger pulse arrives in the midst of a timing period (when Q is high and charging C1 via R1), C1 discharges rapidly via D1 on the low part of the trigger, and commences a new timing cycle as the input waveform switches high again. In practice, the input trigger pulse must be wide enough to fully discharge C1, but should be narrow relative to the width of the output pulse. The output pulse timing period equals $0.7 \times C1 \times R1$. For best results, R1 should have as large a value as possible.

TTL Monostable IC Circuits

In TTL circuitry, the best way of generating high quality pulses is via a dedicated TTL pulse-generator IC such as the 74121, 74LS123, or 74LS221. *Figure 8.31* shows the functional diagram, etc., of the 74121. This old but very popular 'standard' monostable pulse generator IC can give useful output pulse widths from 30nS to hundreds of mS via (usually) two external timing components, and can be configured to give either level-

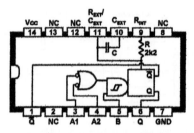

Figure 8.31 Functional diagram, etc., of the 74121 'standard' triggered monostable IC

sensitive rising-edge or simple falling-edge triggering action. Note that the IC has three available trigger-input terminals; of these, A1 and A2 are used as falling-edge triggering inputs, and B functions as a level-sensitive Schmitt rising-edge triggering input; *Figure 8.32* shows how to connect these inputs for a specific type of trigger action.

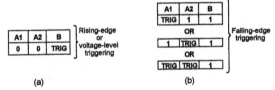

(a) (b)

Figure 8.32 Basic 74121 connections for *(a)* rising-edge or *(b)* falling-edge triggering

Thus, for rising-edge or voltage level triggering, A1 and A2 must be grounded and the trigger input is applied to B, as shown in *Figure 8.33* (which also shows the two external timing components wired in place). For falling-

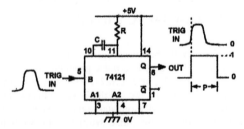

Figure 8.33 Basic 74121 rising-edge or voltage-level triggering connections

edge triggering, B must be tied to logic-1, and the trigger input must be applied to A1 and/or A2, but the unused 'A' input (if any) must be tied to logic-1; *Figure 8.34* shows an example of one of these options, with B and A2 tied to logic-1, and the trigger input applied to A1.

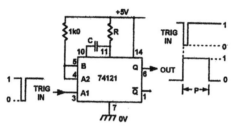

Figure 8.34 One basic set of 74121 falling-edge trigger connections

Dealing next with the 74121's timing circuitry, this IC has three timing-component terminals. A low-value timing capacitor is built into the IC and can be augmented by external capacitors wired between pins 10 and 11 (on polarized capacitors the '+' terminal must go to pin-11). The IC also incorporates a 2k0 timing resistor that is used by connecting pin-9 to pin-14 either directly or via a resistance of up to 40k; alternatively, the internal resistor can be ignored and an external resistance (1k4 to 40k) can be wired between pin-11 and pin-14. Whichever connection is used, the output pulse width = $0.7R_TC_T$, where width is in milliseconds, R_T is the *total* timing resistance, and C_T is the timing capacitance in microfarads. Note that the circuits of *Figures 8.33* and *8.34* show the normal methods of using the 74121, with external timing resistors and capacitors.

Figure 8.35 shows the 74121 used as a 30nS pulse generator, using only the IC's internal timing components, and with its trigger signal connected to the B (Schmitt) input via a simple transistor buffer stage, and *Figure 8.36* shows it used to make an add-on pulse generator that can be used with an existing squarewave 'trigger' generator and spans the range 100ns to 100ms in six decade ranges, using both internal and external timing resistors and decade-switched external capacitors.

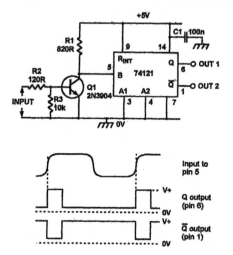

Figure 8.35 30ns pulse generator using 'B' input (Schmitt) rising-edge or voltage-level triggering

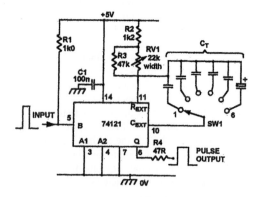

SW1 RANGE	C$_T$ VALUE	PULSE-WIDTH RANGE
1	100pF	100nS - 1µS
2	1n0	1µS - 10µS
3	10n	0µS - 100µS
4	100n	100µS - 1mS
5	1µ0	1mS - 10mS
6	10µ	10mS - 100mS

Figure 8.36 This high-performance add-on pulse generator spans 100ns to 100mS

One type of pulse generator widely used in laboratory work is the delayed-pulse generator, which starts to generate its output pulse at some specified delay-time *after* the application of the initial trigger signal. *Figure 8.37(a)* shows the typical waveforms of this type of generator, and *(b)* shows the basic way of making such a generator from two 74121 ICs. Here, the trigger input signal fires IC1, which generates a 'delay' pulse, and as this pulse ends its inverted (NOT-Q) output fires IC2, which generates the final (delayed) output pulse.

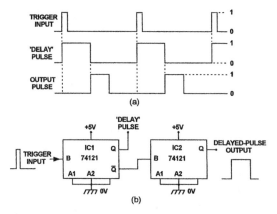

Figure 8.37   Basic delayed-pulse generator waveforms and circuit

Figure 8.38 shows how two of the basic *Figure 8.36* circuits can be coupled together to make a practical 'add-on' wide-range delayed-pulse generator in which both the 'delay' and 'output pulse' periods are fully variable from 100ns to 100ms. Note in this circuit that both fixed-amplitude inverted and non-inverted outputs are short-circuit protected via 47R series resistors. This circuit's timing periods and C_T values are identical to those listed in the table of *Figure 8.36*.

Figure 8.39 shows the functional diagram of the 74LS123 Dual retriggerable monostable with CLEAR, together with basic connections for its external timing components (R and C); note that pins 6 and 14 are internally connected to pin-8 (ground), and that internal capacitances of about 20pF exist between pins 6–7 and

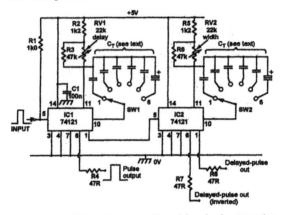

Figure 8.38 High-performance add-on delayed-pulse generator spans 100ns to 100ms

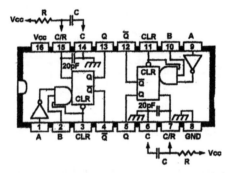

Figure 8.39 Functional diagram and basic external timing-component connections of the 74LS123 Dual retriggerable monostable with CLEAR

pins 14–15. In this IC, each mono has three input trigger-control terminals, notated A, B, and CLR. Normally, CLR must be biased to logic-1 (via a 1k0 resistor); it sets the Q output at logic-0 when CLR is pulled low.

The 74LS123 can be triggered in three different modes via the A, B, and CLR terminals, as shown in *Figures 8.40* and *8.41*. It can be triggered in the rising-edge mode by tying A to logic-0, CLR to logic-1, and applying the trigger signal to B, as shown in *Figure 8.41(a)*, or in the falling-edge mode by tying B and CLR

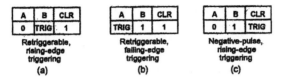

A	B	CLR
0	TRIG	1

Retriggerable,
rising-edge
triggering
(a)

A	B	CLR
TRIG	1	1

Retriggerable,
falling-edge
triggering
(b)

A	B	CLR
0	1	TRIG

Negative-pulse,
rising-edge
triggering
(c)

Figure 8.40 Basic 74LS123 connections for three different triggering modes

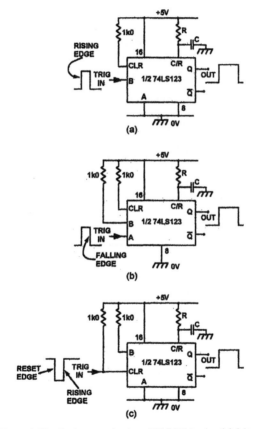

Figure 8.41 Basic ways of using a 74LS123 to give *(a)* rising-edge, *(b)* falling-edge, or *(c)* negative-pulse rising-edge triggering

to logic-1 and applying the trigger signal to A, as shown in *(b)*. The third 'negative-pulse rising-edge' mode – shown in *(c)* – is rarely used; here, A is set to logic-0, B and CLR are biased to logic-1, and the negative-going trigger pulse is applied to CLR; the falling-edge of this pulse resets the monostable, and the rising-edge triggers a new monostable period.

Dealing next with the 74LS123's timing network, this consists of an external resistance (5k0 to 260k) connected between the C/R terminal and the positive supply line, and a capacitance (of any appropriate value) connected between C/R and either the 'C' terminal or ground; *Figure 8.42* shows three alternative ways of connecting these components. The 74LS123's timing period is given approximately by:

$$p = 0.4R_T C_T$$

where 'p' = pulse width in nanoseconds, R_T = resistance in kilohms, and C_T = *total* capacitance (including the internal 20pF) in pF; thus, external values of 25k and 10nF give pulse widths of about 100µs, etc. Note that these values are subject to some variation between individual ICs, and with variations in supply voltage and temperature (typically up to ±1% over the IC's voltage range, and up to –2% over its operating temperature range).

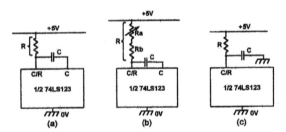

Figure 8.42 Alternative ways of connecting 74LS123 timing components

Before leaving the 74LS123, note that if only one of its monostables is needed, the unwanted monostable should be disabled by tying its CLR terminal high and grounding its A and B terminals, as shown in *Figure 8.43*.

Finally, *Figure 8.44* shows the functional diagram of the 74LS221 Dual precision Schmitt-triggered monostable

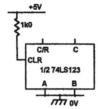

Figure 8.43 Method of disabling an unwanted 74LS123 monostable

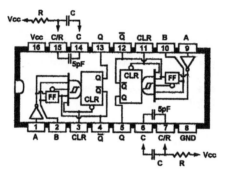

Figure 8.44 Functional diagram and basic external timing-component connections of the 74LS221 Dual precision Schmitt-triggered monostable with CLEAR

with CLEAR, together with basic connections for its external timing components (R and C); note that the external C must be connected between pins 6–7 or 14–15, and that these pins are shunted by an internal capacitance of about 5pF. Also note that the 74LS221 is NOT a retriggerable IC, and is thus subject to the normal duty-cycle limitations of an ordinary monostable; the IC should in fact be regarded as an improved high-performance Dual version of the old 74121 (at the time of writing, it costs less than half the 74121 price). In the 74LS221, each mono has three Schmitt-type input trigger-control terminals, notated A, B, and CLR, with semi-latching action on B and CLR via an internal NAND-type flip-flop (FF). Normally, CLR must be biased to logic-1 (via a 1k0 resistor); it sets the Q output at logic-0 if it is pulled low.

The 74LS221 can be triggered in three different modes via the A, B, and CLR terminals, as shown in *Figures*

8.45 and *8.46*. It can be triggered in the rising-edge mode by tying A to logic-0, CLR to logic-1, and applying the trigger signal to B, as shown in *Figure 8.46(a)*, or in the falling-edge mode by tying B and CLR to logic-1 and applying the trigger signal to A, as shown in *(b)*. The third 'negative-pulse rising-edge' mode is rarely used; here, A is set to logic-0, and the circuit is then primed by switching B from logic-0 to logic-1 while CLR is held at logic-0; then, with B at logic-1, a rising-edge on CLR will trigger a monostable period.

A	B	CLR
0	TRIG	1

Rising-edge
Schmitt
triggering
(a)

A	B	CLR
TRIG	1	1

Falling-edge
Schmitt
triggering
(b)

A	B	CLR
0	1	TRIG

Negative-pulse
rising-edge Schmitt
triggering (see text).
(c)

Figure 8.45   Basic 74LS221 connections for three different triggering modes

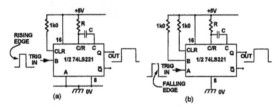

Figure 8.46   Basic ways of using a 74LS221 to give *(a)* rising-edge or *(b)* falling-edge Schmitt triggering action

Dealing next with the 74LS221's timing network, this consists of an external resistance (1k4 to 100k) connected between the C/R terminal and the positive supply line, and a capacitance (of any value up to 1000μF) connected between C/R and the C terminal, as shown in *Figure 8.46*. The 74LS221's timing period is given approximately by:

$$p = 0.7R_TC_T$$

where 'p' = pulse width in nanoseconds, R_T = resistance in kilohms, and C_T = *total* capacitance (including the internal 5pF) in pF; thus, external values of 25k and 10nF give pulse widths of about 175μs, etc. Note that these values are virtually independent of variations in supply voltage and temperature over the IC's full voltage/temperature operating range.

In applications where only one of the 74LS221's monostables is needed, the unwanted monostable should be disabled by tying its CLR terminal high and grounding its A and B terminals, as shown in *Figure 8.47*.

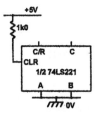

Figure 8.47 Method of disabling an unwanted 74LS221 monostable

'4000'-series Monostable ICs

Several dedicated CMOS monostable ICs are readily available, and the best known '4000'-series versions of these are the 4047B monostable/astable IC, and the 4098B Dual monostable (a greatly improved version of the 4528B). *Figures 8.48* and *8.49* show the functional

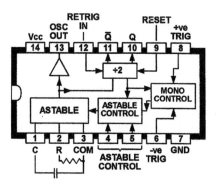

Figure 8.48 Functional diagram and external timing component locations of the 4047B monostable/astable IC

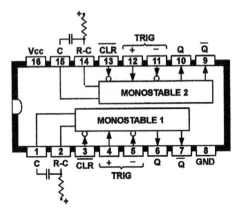

Figure 8.49   Functional diagram and external timing component locations of the 4098B Dual monostable IC

diagrams of these two devices, together with the locations of their external timing components. These ICs give only moderately good pulse-width accuracy and stability, but can be triggered by either the positive or the negative edge of an input signal and can be used in either the standard or the retriggerable mode.

The 4047B houses an astable multi and a divide-by-2 stage, plus logic networks. When used in the monostable mode the trigger signal starts the astable and resets the counter, driving its Q output high; after two C–R-controlled astable cycles the counter flips over and simultaneously kills the astable and switches the Q output low, completing the operating sequence. The C–R timing components produce relatively long output pulse periods, which approximate 2.5CR (R can have any value from 10k to 10M, and C must be a non-polarized capacitor with a value greater than 1nF). *Figures 8.50(a)* and *8.50(b)* show how to connect the IC as a standard monostable triggered by either positive *(a)* or negative *(b)* input edges, and *Figure 8.50(c)* shows how to connect the monostable in the retriggerable mode; these circuits can be reset at any time by pulling RESET pin-9 high.

The 4047B can be used as a free-running astable multivibrator (squarewave generator) by connecting it as in *Figure 8.51*. The user has two output options in this

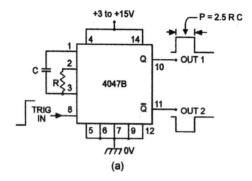

(a)

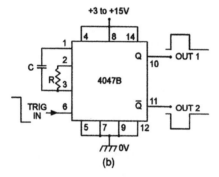

(b)

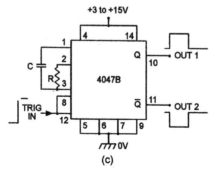

(c)

Figure 8.50 Various ways of using the 4047B as a monostable. *(a)* Positive-edge-triggered monostable. *(b)* Negative-edge-triggered monostable. *(c)* Retriggerable monostable, positive-edge triggered

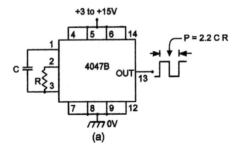

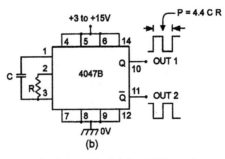

Figure 8.51   Basic ways of using the 4047B as a free-running astable multivibrator. *(a)* Direct squarewave output (may not be symmetrical). *(b)* Divide-by-2 outputs are perfectly symmetrical

mode. Pin-13 allows the output of the internal astable to be directly accessed; this output may not be precisely symmetrical, and has a period of 2.2CR. Alternatively, the output may be taken (from pin-10 or 11) via the internal divide-by-2 element, in which case the output is perfectly symmetrical and has a period of 4.4CR.

Figure 8.52 shows various ways of using the 4098B Dual monostable IC. This is a fairly simple device, in which the two monostable sections share common supply connections but can otherwise be used independently. Mono-1 is housed on the left side (pins 1 to 7) of the IC, and mono-2 on the right side (pins 9 to 15). The timing period of each mono is controlled by a single resistor (R) and capacitor (C), and approximates 0.5CR. R can have any value from 5k0 to 10M, and C can have any value from 20pF to 100µF. Note in the diagrams of *Figure 8.52* that the bracketed numbers relate to the pin connections of mono-2, and the plain numbers to mono-1, and that the RESET terminal (pins 3 or 13) is shown disabled.

Figures 8.52(a) and *8.52(b)* show how to use the 4098B to make retriggerable monostables that are triggered by positive or negative input edges respectively. In *Figure 8.52(a)* the trigger signal is fed to the '+' TRIG pin, and the '–' TRIG pin is tied low. In *Figure 8.52(b)* the trigger signal is applied to the '–' TRIG pin, and the '+' TRIG pin is tied high.

Figures 8.52(c) and *8.52(d)* show how to use the IC to make standard (non-retriggerable) monostables that are

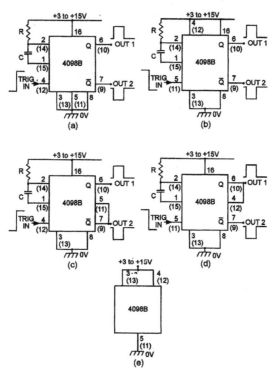

Figure 8.52 Various ways of using the 4098B monostable. *(a)* Positive-edge-triggering, retriggerable mono. *(b)* Negative-edge-triggering, retriggerable mono. *(c)* Positive-edge-triggering (non-retriggerable mono). *(d)* Negative-edge-triggering (non-retriggerable mono). *(e)* Connections for each unused section of the IC

triggered by positive or negative edges respectively. These circuits are similar to those mentioned above except that the unused trigger pin is coupled to either the Q or the NOT-Q output, so that trigger pulses are blocked once a timing cycle has been initiated.

Finally, *Figure 8.52(e)* shows how the unused half of the IC must be connected when only a single monostable is wanted from the package. The '−' TRIG pin is tied low, and the '+' TRIG and RESET pins are tied high.

'74HC'-series Monostable ICs

The three best known '74HC'-series monostable ICs are the 74HC123, 74HC221, and the 74HC4538. These are all Dual monostable types. The 74HC4538 is a Dual retriggerable mono, with CLEAR, and is functionally almost identical to the 4098B, and uses the same pin notations, etc. (see *Figure 8.49*); each monostable generates a pulse with a period (width) of 0.7CR.

The 74HC123 (see *Figure 8.53*) is a Dual retriggerable monostable, with CLEAR, and is a modified fast CMOS version (with Schmitt 'trigger' inputs) of the popular 74LS123 TTL IC, but its 'C' terminals (pins 6 and 14) are not internally connected to the GND terminal, and in

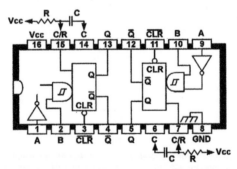

Figure 8.53 Functional diagram and external timing component locations of the 74HC123 Dual retriggerable monostable with CLEAR. *Note*: In use, terminal C (pins 6 and 14) must be hard-wired to GND

use these pins must be hard-wired to GND. The IC is used in a similar way to the 74LS123 (see _Figures 8.54 to 8.55_), but each monostable generates an output pulse with a period of $1.0 \times CR$, and R can have any value up to 10M.

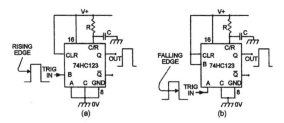

Figure 8.54 Basic ways of using the 74HC123 to give _(a)_ rising-edge or _(b)_ falling-edge triggering

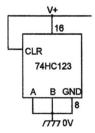

Figure 8.55 Ways of disabling an unwanted 74HC123 monostable element

Finally, the 74HC221 is another Dual retriggerable monostable with CLEAR and with Schmitt 'trigger' inputs, and is simply a modified fast CMOS version of the popular 74LS221 TTL IC. It is, for most practical purposes, functionally similar to the 74HC123, and is used in the ways already shown in _Figures 8.54_ and _8.55_. Like the 74HC123, its 'C' terminals (pins 6 and 14) are not internally connected to the GND terminal, and in use must be hard-wired to GND. Each monostable generates an output pulse with a period of $1.0 \times CR$, and R can have any value up to 10M.

9 Squarewave Generator Circuits

Squarewave generators are among the most widely used of all circuits used in modern electronics. They can be used for flashing LED indicators, for generating audio and alarm tones, or, if their leading and trailing edges are really sharp, for 'clocking' logic or counter/divider circuitry, etc. Such circuits can be designed to give symmetrical or non-symmetrical outputs, and can be of the free-running or the gated types; in the latter case, they can be designed to turn on with either logic-0 or logic-1 gate signals, and to give either a logic-0 or a logic-1 output when in the OFF mode. The designs can be based on a variety of semiconductor technologies, including the humble transistor, the op-amp, the 555-timer IC, or on CMOS or TTL logic elements, etc. This chapter looks at a variety of designs based on popular TTL and CMOS ICs.

TTL Schmitt Astable Circuits

In TTL applications, the easiest and most cost-effective way to make a squarewave generator is by using a 74LS14 or similar Schmitt inverter element in the basic astable configuration shown in the circuit of *Figure 9.1*, which operates as follows.

Suppose in *Figure 9.1* that C's voltage has just fallen to the Schmitt's lower threshold value of 0.8V, making the

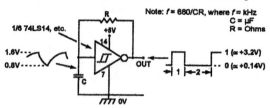

Figure 9.1 Basic circuit and waveforms, etc., of a TTL Schmitt astable

Schmitt's output switch to logic-1; under this condition the Schmitt output is at about +3.2V, so C starts to charge exponentially upwards from 0.8V until it reaches the 1.6V upper threshold value of the Schmitt, at which point the Schmitt's output switches abruptly to a logic-0 value of about 0.14V, and C starts to discharge exponentially downwards from 1.6V until it reaches the 0.8V lower threshold value, at which point the Schmitt's output switches to logic-1 again, and the whole process repeats again, and so on.

This simple TTL Schmitt astable circuit generates a useful but non-symmetrical squarewave output; its Mark–Space ratio is about 1:2 (i.e. it has a 33% duty cycle), and its operating frequency (*f*) approximately equals 680/(CR), where C is in µF, R (which can have any value in the 100R to 1k2 range) is in ohms, and *f* is in kHz; thus, C and R values of 100nF and 1k0 give an operating frequency about 6.8kHz, etc. Note that the operating frequency has a slight positive temperature coefficient, and has a supply voltage coefficient of about +0.5%/100mV. The circuit can, in theory, operate at frequencies ranging from below 1Hz (C = 1000µF) to above 10MHz (C = 50pF), but in practice is best limited to the approximate frequency range 400Hz – 2MHz, because the need for large C values makes it uneconomic at low frequencies, and it has poor stability at high frequencies.

The basic TTL Schmitt astable circuit can be usefully modified in a variety of ways. The operating frequency can, for example, be made variable by using a fixed 100R and variable 1k0 resistor in the R position, and the output waveform can be improved by feeding it through a Schmitt buffer stage, as shown in *Figure 9.2*. If perfect

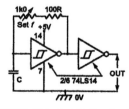

Figure 9.2 Variable-frequency buffered-output TTL Schmitt astable

waveform symmetry is needed, it can be obtained by feeding the output of a buffered Schmitt astable through a JK flip-flop, as shown in *Figure 9.3*, but note that the final output frequency is half that of the astable. The

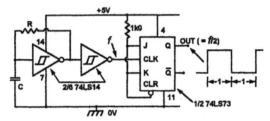

Figure 9.3 TTL Schmitt astable with precise 1:1 Mark–Space ratio output

basic circuit can be converted into a gated Schmitt astable by using a 74LS132 2-input Schmitt NAND gate as its basic element, as shown in *Figure 9.4*; this particular circuit is gated on by a logic-1 input and has a

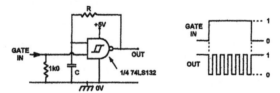

Figure 9.4 Gated TTL Schmitt astable with logic-1 gate on and normally high output

normally high (logic-1) output; it can be made to give a normally low (logic-0) output by feeding the output through a spare 74LS132 element connected as a simple Schmitt inverter, as shown in *Figure 9.5*.

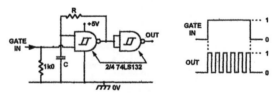

Figure 9.5 Gated TTL Schmitt astable with logic-1 gate on and normally low output

A major drawback of the TTL Schmitt astable is that its low maximum value of timing resistor (1k2) makes it necessary to use very large values of timing capacitor at low operating frequencies. At 1kHz, for example, timing component values of 1uF and 680R are needed. One alternative way of making a 1kHz TTL squarewave generator is to use a single 10nF capacitor to make a precision 100kHz astable, and then divide its output frequency by 100 via two decade counter ICs, as shown in the circuit of *Figure 9.6*, which provides outputs of 100kHz, 10kHz, and 1kHz.

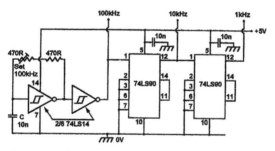

Figure 9.6   This expanded 100kHz TTL Schmitt astable circuit gives 10kHz and 1kHz outputs

CMOS Schmitt Astable Circuits

CMOS Schmitt astables have two major advantages over TTL types. The first is that, because CMOS offers a very high input impedance, they can use large values of timing resistor (up to 10M) and low values of 'C' to set a given frequency. The second is that, because CMOS Schmitt elements have reasonably symmetrical upper and lower trigger threshold values, they generate reasonably symmetrical squarewave outputs. Suitable CMOS ICs for use in this type of application are the 40106B Hex Schmitt inverter (see *Figure 6.28*) and the 4093B Quad 2-input NAND Schmitt trigger (see *Figure 6.53*). In the latter case, each NAND gate of the 4093B can be used as an inverter by simply disabling one of its input terminals, as shown in the basic Schmitt astable circuit of *Figure 9.7*.

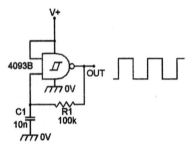

Figure 9.7   Basic CMOS Schmitt astable

The Schmitt astable circuit gives an excellent performance, with very clean output edges that are unaffected by supply-line ripple and other nasties. The operating frequency is determined by the C1–R1 values, and can be varied from a few cycles per minute to 1MHz or so, the upper limit being determined by the practical limitations of real-life timing resistors and capacitors, etc. The circuit action is such that C1 alternately charges and discharges via R1, without switching the C1 polarity; C1 can thus be a non-polarized component. Note that fast '74HC'-series Schmitt elements generate sharper squarewave edges than '4000'-series types, but otherwise offer little real advantage over the latter type.

Figure 9.8 shows how the above 4093B-based astable circuit can be modified so that it can be gated on and off via an external signal; the circuit is gated on by a high (logic-1) input, but gives a high output when it is in the gated-off state.

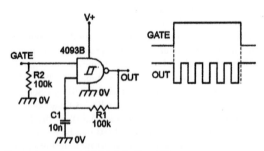

Figure 9.8   Gated CMOS Schmitt astable

The basic *Figure 9.7* astable circuit generates an inherently symmetrical squarewave output. The basic circuit can be made to produce a non-symmetrical output by providing its timing capacitor with alternate charge and discharge paths, as shown in the designs of *Figures 9.9* and *9.10*. The output M–S ratio of the *Figure 9.9* circuit is fixed, but that of the *Figure 9.10* circuit can be varied over a wide range via RV1.

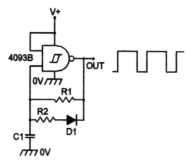

Figure 9.9   CMOS astable with non-symmetrical M–S ratio

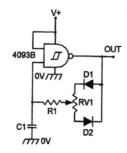

Figure 9.10   CMOS astable with variable M–S ratio

2-stage CMOS Astable Basics

One very popular way to make a CMOS squarewave generator is to wire two CMOS inverter stages in series and use the C–R feedback network shown in the basic 2-stage astable multivibrator circuit of *Figure 9.11(a)*. This

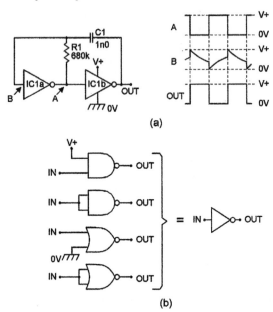

Figure 9.11   *(a)* Circuit and waveforms of basic 2-stage 1kHz CMOS astable. *(b)* Ways of connecting a 2-input NAND or NOR gate as an inverter

circuit generates a good squarewave output from IC1b (and a less-good anti-phase squarewave output from IC1a), and operates at about 1kHz with the component values shown. The circuit operates as follows.

In *Figure 9.11*(a) the two inverters are wired in series, so the output of one goes high when the other goes low, and vice versa. Time-constant network C1–R1 is wired between the outputs of IC1b and IC1a, with the C1–R1 junction fed to the input of the IC1a inverter stage. Suppose initially that C1 is fully discharged, and that the output of IC1b has just switched high (and the output of IC1a has just switched low).

Under this condition, the C1–R1 junction voltage is initially at full positive supply volts, driving the output of IC1a hard low, but this voltage immediately starts to decay exponentially as C1 charges up via R1, until eventually it falls into the linear transfer voltage range of IC1a, making its output start to swing high. This swing is amplified by inverter IC1b, initiating a regenerative action in which IC1b output switches abruptly to the low state (and IC1a output switches high). This switching action makes the charge of C1 try to apply a negative voltage to the input of IC1a, but the built-in protection diodes of IC1a prevent this, and instead discharge C1.

Thus, at the start of the second cycle, C1 is again fully discharged, so in this case the C1–R1 junction is initially at zero volts (driving IC1a output high), but the voltage then rises exponentially as C1 charges up via R1, until eventually it rises into the linear transfer voltage range of IC1a, thus initiating another regenerative switching action in which IC1b output switches high again (and IC1a output switches low), and C1 is initially discharged via the IC1a input protection diodes. The operating cycle then continues *ad infinitum*.

The operating frequency of the above circuit is inversely proportional to the C–R time constant (the period is roughly 1.4CR), so can be raised by lowering the values of C1 or R1. C1 must be a non-polarized capacitor and can have any value from a few tens of pF to several μF, and R1 can have any value from about 4k7 to 22M; the astable operating frequency can vary from a fraction of a Hz to about 1MHz. For variable-frequency operation, wire a fixed and a variable resistor in series in the R1 position.

In practice, each of the circuit's inverter stages can be made from a single gate of a 4001B Quad 2-input NOR gate or a 4011B Quad 2-input NAND gate, etc., by using the connections shown in *Figure 9.11(b)*; the inputs of all unused gates in these ICs must be tied to one or other of the supply line terminals. The *Figure 9.11*(a) astable can (like all other CMOS astable circuits shown in this chapter) use any supplies in the range 3V to 15V if based on a '4000'-series IC such as the 4001B or 4011B, etc., or 2V to 6V if based on a '74HC'-series device.

The output of the *Figure 9.11(a)* astable switches (when lightly loaded) almost fully between the zero and positive supply-rail values, but the C1–R1 junction voltage is prevented from swinging below zero or above the positive supply-rail levels by the built-in clamping diodes at the input of IC1a. This factor makes the operating frequency somewhat dependent on supply-rail voltages. Typically, the frequency falls by about 0.8% for a 10% rise in supply voltage; if the frequency is normalized with a 10V supply, the frequency falls by 4% at 15V or rises by 8% at 5V.

The operating frequency of the *Figure 9.11(a)* circuit is also influenced by the transfer voltage value of the individual IC1a inverter/gate that is used in the astable, and can be expected to vary by as much as 10% between different ICs. The output symmetry of the 'square' waveform also depends on the transfer voltage value, and in most cases the circuit will give a non-symmetrical output. In most non-precision and hobby applications these defects are, however, of little practical importance.

2-stage Astable Variations

Some of the defects of the *Figure 9.11(a)* circuit can be minimized by using the 'compensated' astable of *Figure 9.12*, in which R2 is wired in series with the input of IC1a. This resistor must have a value that is large relative to R1, and its main purpose is to allow the C1–R1 junction to swing freely below the zero and above the positive supply-rail voltages during the astable operation and thus improve the frequency stability of the circuit. Typically, when R2 is ten times the value of R1, the frequency varies by only 0.5% when the supply voltage is varied between 5 and 15 volts. An incidental benefit of R2 is that it gives a slight improvement in the symmetry of the astable output waveform.

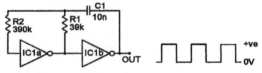

Figure 9.12 This 'compensated' version of the 1kHz astable has excellent frequency stability

The basic and compensated astable circuits of *Figures 9.11* and *9.12* can be built with several detail variations, as shown in *Figures 9.13* to *9.17*. In the basic astable

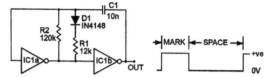

Figure 9.13 Modifying the astable to give a non-symmetrical output: MARK is controlled by the parallel values of R1 and R2: SPACE is controlled by R2 only

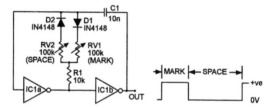

Figure 9.14 This astable has independently variable MARK and SPACE times

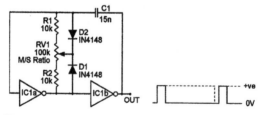

Figure 9.15 The MARK–SPACE ratio of this astable is fully variable from 1:11 to 11:1 via RV1; frequency is almost constant at about 1kHz

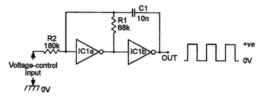

Figure 9.16 Simple VCO circuit

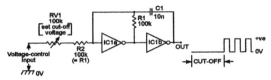

Figure 9.17   Special-effects VCO which cuts off when V$_{in}$ falls below a preset value

circuit, for example, C1 alternately charges and discharges via R1 and thus has a fixed symmetry. *Figures 9.13* to *9.15* show how the basic circuit can be modified to give alternate C1 charge and discharge paths and thus to allow the symmetry to be varied at will.

The *Figure 9.13* circuit generates a highly non-symmetrical waveform, equivalent to a fixed pulse delivered at a fixed timebase rate. Here, C1 charges in one direction via R2 in parallel with the D1–R1 combination, to generate the MARK or pulse part of the waveform, but discharges in the reverse direction via R2 only, to give the SPACE between the pulses. The *Figure 9.14* circuit, however, generates a waveform with independently variable MARK and SPACE times; the MARK is controlled by R1-RV1-D1, and the SPACE by R1–RV2–D2. Finally, the *Figure 9.15* circuit generates a variable symmetry or M–S ratio output while maintaining a near-constant frequency; here, C1 charges in one direction via D2 and the lower half of RV1 and R2, and in the other direction via D1 and the upper half of RV1 and R1, and the M–S ratio can be varied over the range 1:11 to 11:1 via RV1.

Figures 9.16 and *9.17* show ways of using the basic 2-stage astable circuit as a very simple VCO. The *Figure 9.16* circuit can be used to vary the operating frequency over a limited range via an external voltage. R2 must be at least twice as large as R1 for satisfactory operation, the actual value depending on the required frequency-shift range: a low R2 value gives a large shift range, and a large R2 value gives a small shift range. The *Figure 9.17* circuit acts as a special-effects VCO in which the oscillator frequency rises with input voltage, but switches off completely when the input voltage falls below a value preset by RV1.

Gated 2-stage Astable Circuits

All of the 2-stage CMOS astable circuits of *Figures 9.11*
to *9.15* can be modified for gated operation, so that they
can be turned on and off via an external signal, by
simply using a 2-input NAND (4011B, etc.) or NOR
(4001B, etc.) gate in place of the inverter in the IC1a
position and by applying the input gate control signal to
one of the gate input terminals. Note, however, that the
NAND and the NOR elements give quite different types
of gate control and output operation in these
applications, as shown by the two basic versions of the
gated astable in *Figures 9.18* and *9.19.*

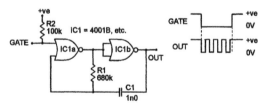

Figure 9.18 This gated astable has a normally low output and is
gated on by a high (logic-1) input

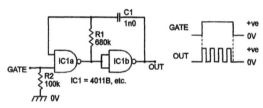

Figure 9.19 This version of the gated astable has a normally
high output and is gated on by a low (logic-0) input

Note specifically from these two circuits that the NAND
version is gated on by a logic-1 input and has a normally
low output, while the NOR version is gated on by a
logic-0 input and has a normally high output. R2 can be
eliminated from these circuits if the gate drive is direct-
coupled from the output of a preceding CMOS logic

stage, etc. Also note in these gated astable circuits that the output signal terminates as soon as the gate drive is removed; consequently, any noise present at the gate terminal also appears at the outputs of these circuits. *Figures 9.20* and *9.21* show how to modify the circuits so that they produce 'noiseless' outputs.

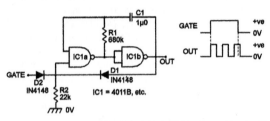

Figure 9.20 Semi-latching or 'noiseless' gated astable circuit, with logic-1 gate input and normally zero output

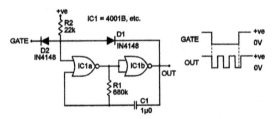

Figure 9.21 Alternative semi-latching gated astable, with logic-0 gate input and normally high output

Here, the gate signal of IC1a is derived from both the outside world and from the output of IC1b via diode OR gate D1–D2–R2. As soon as the circuit is gated from the outside world via D2 the output of IC1b reinforces or self-latches the gating via D1 for the duration of one half astable cycle, thus eliminating any effects of a noisy outside world signal. The outputs of these semi-latching gated 2-stage astable circuits are thus always complete numbers of half cycles.

CMOS Ring-of-three Astable

The CMOS 2-stage astable circuit is a good general-purpose squarewave generator, but is not always suitable for direct use as a 'clock' generator with fast-acting counting and dividing circuits, since it tends to pick up and amplify any existing supply line noise during the 'transitioning' parts of its operating cycle and thus to produce output squarewaves with 'glitchy' leading and trailing edges. A far better type of clock generator circuit is the ring-of-three astable shown in *Figure 9.22*.

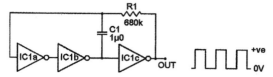

Figure 9.22 This ring-of-three astable makes an excellent clock generator

The *Figure 9.22* ring-of-three circuit is similar to the basic 2-stage astable, except that its input stage (IC1a–IC1b) acts as an ultra-high-gain non-inverting amplifier and its main timing components (C1–R1) are transposed (relative to the 2-stage astable). Because of the very high overall gain of the circuit, it produces an excellent and glitch-free squarewave output that is ideal for clock generator use.

The basic ring-of-three astable can be subjected to all the design modifications already described for the basic 2-stage astable, e.g. it can be used in either basic or compensated form and can give either a symmetrical or non-symmetrical output, etc. The most interesting variations of the circuit occur, however, when it is used in the gated mode, since it can be gated via either the IC1b or IC1c stages. *Figures 9.23* to *9.26* show four variations on this gating theme.

Thus, the *Figure 9.23* and *9.24* circuits are both gated on by a logic-1 input signal, but the *Figure 9.23* circuit has a normally low output, while that of *Figure 9.24* is

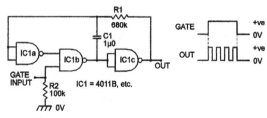

Figure 9.23   This gated ring-of-three astable is gated by a logic-1 input and has a normally low output

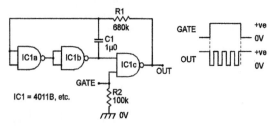

Figure 9.24   This gated ring-of-three astable is gated by a logic-1 input and has a normally high output

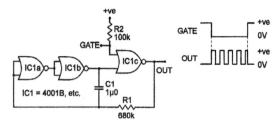

Figure 9.25   This gated ring-of-three astable is gated by a logic-0 input and has a normally low output

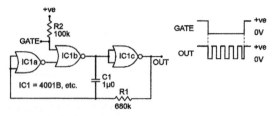

Figure 9.26   This gated ring-of-three astable is gated by a logic-0 input and has a normally-high output

normally high. Similarly, the *Figure 9.25* and *9.26* circuits are both gated on by a logic-0 signal, but the output of the *Figure 9.25* circuit is normally low, while that of *Figure 9.26* is normally high.

CMOS 555 IC Circuits

Most readers will know that the popular 555 timer IC can be wired in the astable mode and used to generate excellent squarewave output signals, and that a CMOS version of this device is also available and is known as the ICM7555 or – more commonly – as the '7555'. *Figure 9.27* shows the outline of this CMOS IC, and *Figure 9.28* shows it wired in the basic astable mode.

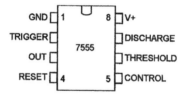

Figure 9.27 Outline of the 7555 'CMOS 555' IC

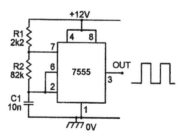

Figure 9.28 Basic 7555 astable

The action of the *Figure 9.28* basic 7555 astable is such that C1 alternately charges via R1–R2 and discharges via R2 only, generating an excellent near-symmetrical squarewave output that is suitable for use as a clock waveform. Note that R2 acts as the main time-constant resistor, and can have any value from about 4k0 to 22M, and C1 is the time-constant capacitor, and can have any value from a few pF to many hundreds of μF, and may be of the polarized or non-polarized types.

The 7555 astable circuit can only work if pin-4 is tied to the positive supply rail; if pin-4 is grounded, the astable is disabled. The astable can thus be used in the gated mode by simply wiring pin-4 as shown in *Figure 9.29*.

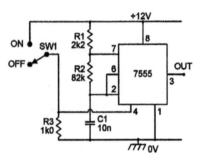

Figure 9.29   Gated 7555 astable

The basic 7555 astable circuit generates an almost symmetrical output waveform. It can be made to generate a non-symmetrical waveform in a variety of ways. *Figure 9.30* shows one useful variation. In this case C1 alternately charges via R1–R3 and D2, and discharges via D1–RV1 and R2; the output waveform symmetry of this circuit is thus fully variable via RV1.

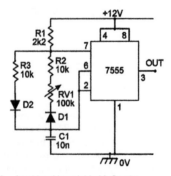

Figure 9.30   Astable with variable M–S ratio

4046B VCO Circuits

Another popular 'CMOS' way of generating good squarewaves is via the VCO (voltage-controlled oscillator) section of the 4046B phase-locked-loop (PLL) IC (or its 74HC-series equivalent, the 74HC4046). *Figure 9.31* shows the internal block diagram and pin-outs of this excellent IC, which contains a couple of phase comparators, a VCO, a zener diode, and a few other bits and pieces.

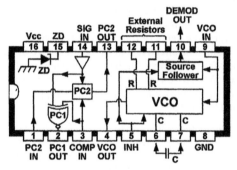

Figure 9.31 Internal block diagram and pin-outs of the 4046B (and its 'HC' equivalent, the 74HC4046)

For present purposes, the most important part of the IC is the VCO section. This is a highly versatile element. It produces a well-shaped symmetrical squarewave output, has a top-end frequency limit in excess of 1MHz (15MHz in the 74HC4046), has a voltage-to-frequency linearity of about 1% and can easily be scanned through a 1 000 000:1 range by an external voltage applied to the VCO input terminal. The frequency of the oscillator is governed by the value of a capacitor (minimum value 50pF) connected between pins 6 and 7, by the value of a resistor (minimum value 10k) wired between pin-11 and ground, and by the voltage (any value from zero to the supply voltage value) applied to VCO input pin-9.

Figure 9.32 shows the simplest possible way of using the 4046B VCO as a voltage-controlled squarewave generator. Here, C1–R1 determine the maximum

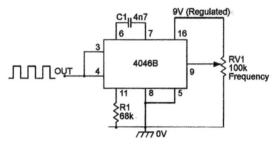

Figure 9.32   Basic wide-range VCO, spanning near zero to roughly 5kHz via RV1

frequency that can be obtained (with the pin-9 voltage at maximum) and RV1 controls the actual frequency by applying a control voltage to pin-9; the frequency falls to a very low value (a fraction of a Hz) with pin-9 at zero volts. The effective voltage-control range of pin-9 varies from roughly 1V0 below the supply value to about 1V0 above zero, and gives a frequency span of about 1 000 000:1. Ideally, the circuit's supply voltage should be regulated.

It is stated above that the frequency falls to near-zero when the input voltage of the *Figure 9.32* circuit is reduced to zero. *Figure 9.33* shows how the circuit can be modified so that the frequency falls all the way to zero with zero input, by wiring high-value resistor R2 between pins 12 and 16. Note here that, when the frequency is reduced to zero, the VCO output randomly settles in either a logic-0 or a logic-1 state. *Figure 9.34*

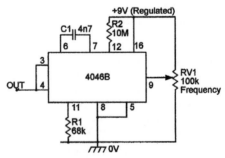

Figure 9.33   The frequency of this VCO is variable all the way down to zero

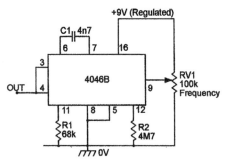

Figure 9.34 Restricted-range VCO, with frequency variable from roughly 72Hz to 5kHz via RV1

shows how the pin-12 resistor can alternatively be used to determine the minimum operating frequency of a restricted-range VCO. Here, f_{min} is determined by C1–R2, and f_{max} is determined by C1 and the parallel resistance of R1 and R2.

Figure 9.35 shows an alternative version of the restricted-range VCO, in which f_{max} is controlled by C1–R1, and f_{min} is determined by C1 and the series combination of R1 and R2. Note that, by suitable choice of the R1 and R2 values, the circuit can be made to span any desired frequency range from 1:1 to near infinity.

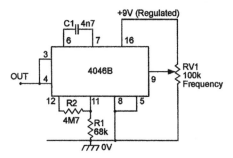

Figure 9.35 Alternative version of the restricted-range VCO. f_{max} is controlled by C1-R1, f_{min} by C1 – (R1 + R2)

Finally, it should be noted that the VCO section of the 4046B can be disabled by taking pin-5 of the package high (to logic-1) or enabled by taking pin-5 low (to logic-0). This feature makes it possible to gate the VCO on and off via external signals. Thus, *Figure 9.36* shows how the basic VCO circuit can be gated via a signal applied to an external inverter stage. Alternatively, *Figure 9.37* shows how one of the internal phase comparators of the 4046B can be used to provide gate inversion, so that the VCO can be gated via an external voltage applied to pin-3.

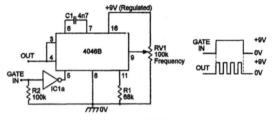

Figure 9.36 Gated wide-range VCO, using an external gate inverter

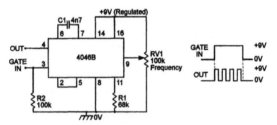

Figure 9.37 Gated wide-range VCO, using one of the internal phase comparators as a gate inverter

A TTL Crystal Oscillator

To complete this look at squarewave generator circuits, *Figure 9.38* shows how two simple 74LS04 or similar TTL inverter elements can be used as the basis of a crystal oscillator by biasing them into their linear modes via 470R feedback resistors and then AC coupling them in series via C1, to give zero overall phase shift; the circuit is then made to oscillate by wiring the crystal

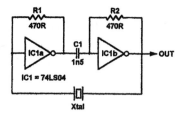

Figure 9.38   TTL crystal oscillator, for use with series-resonant crystals

(which must be a series resonant type) between the output and input as shown. This circuit can operate from a few hundred kHz to above 10MHz.

10　Clocked Flip-flops

Most digital ICs are based on either simple logic gate networks of the types described in earlier chapters, or on clocked bistable or flip-flop elements. This second type includes simple counter/divider ICs, shift registers, data latches, and complex ICs such as presettable up/down counters and dividers. The present chapter explains how clocked flip-flop circuits work, and presents practical user information on some popular clocked flip-flop and counter/divider ICs. As an immediate follow-up, Chapters 11 and 12 look at advanced counter/divider ICs and various associated devices.

Clocked Flip-flop Basics

The simplest type of flip-flop is the cross-coupled bistable, which has already been briefly described in Chapter 8 (see *Figures 8.7* to *8.17*). *Figure 10.1* shows the basic circuit, symbol and Truth Table of a NOR-gate version of this flip-flop, which has two input terminals and a pair of anti-phase output terminals (Q and NOT-Q). The circuit's basic action is such that its Q output latches high (and NOT-Q latches low) when the SET terminal is briefly driven high (to logic-1), and remains in that state until the RESET terminal is briefly driven high, at which point the Q output latches low (and the NOT-Q output latches high). The basic SET-RESET (S–R, or R–S) flip-flop thus acts as a simple memory

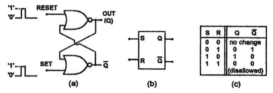

Figure 10.1　Circuit *(a)*, symbol *(b)* and Truth Table *(c)* of a NOR-type S–R flip-flop

element that 'remembers' which of the two inputs last went high. Note that if both inputs go high simultaneously, both outputs go low, but if both inputs then simultaneously switch low the output states cannot be predicted; the 'both inputs high' condition is thus regarded as a disallowed state.

The NAND version of the S–R flip-flop (see *Figure 8.11*) is similar to the NOR type, except that it is triggered by logic-0 inputs. Note at this point that the S–R flip-flop is actually a Schmitt-like voltage-triggered regenerative switch; the NOR type triggers when its input (SET or RESET) voltage *rises* to some intermediate value (between logic-0 and logic-1) at which the CMOS element is biased into its linear amplifying mode; a NAND type triggers when the input *falls* to some intermediate value. Thus, all NOR-type flip-flops are intrinsically 'level-sensitive, rising-edge-triggered' elements, and NAND-type flip-flops are intrinsically 'level-sensitive, falling-edge-triggered' elements; these basic expressions form part of modern flip-flop jargon.

The versatility of the basic *Figure 10.1* circuit can be greatly enhanced by wiring an AND gate in series with each input terminal, using the connections shown in *Figure 10.2*, so that high input signals can only reach the

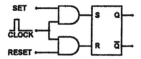

Figure 10.2 Basic circuit of the clocked R–S flip-flop

S–R flip-flop when the clock (CLK) signal is also high. Thus, when the clock signal is low, both inputs of the S–R flip-flop are held low, irrespective of the states of the SET and RESET inputs, and the flip-flop acts as a permanent memory, but when the clock signal is high the circuit acts as a standard R–S flip-flop. Consequently, information is not automatically latched into the flip-flop, but must be clocked in via the CLK terminal: this circuit is thus known as a 'clocked S–R (or R–S) flip-flop'.

The Master–slave Flip-flop

The most important of all basic flop-flop elements is the so-called 'clocked master–slave' type, and *Figure 10.3* shows how one of these can be made from two cascaded S–R flip-flops that are clocked in anti-phase (via an inverter in the clock line). The basic action of this circuit is such that, when the CLK input terminal is low, the inputs to the master flip-flop are enabled via the inverter, so the SET–RESET data is accepted, but the inputs to the slave flip-flop are disabled, so this data is not passed to the output terminals. When the CLK input terminal goes high, however, the inputs to the master flip-flop are disabled via the inverter, which thus outputs only the remembered input data, and simultaneously the input to the slave flip-flop is enabled and the remembered data is latched and passed to the output terminals.

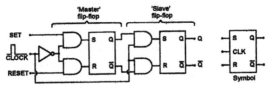

Figure 10.3   Clocked master–slave flip-flop basic circuit and symbol

Thus, the clocked master–slave flip-flop accepts input data or information only when the clock signal is low, and passes that data to the output on the arrival of the rising edge of the clock signal, i.e. its data-shifting action is synchronous with the timing of the clock signal. The clocked master–slave flip-flop is such an important device that it is given its own circuit symbol, as shown in the diagram.

The clocked master–slave flip-flop can be made to give a divide-by-2 'toggle' action by cross-coupling its input and output terminals as shown in *Figure 10.4*, so that SET and Q (and RESET and NOT-Q) logic levels are always opposite. Consequently, when the clock signal is low the master flip-flop receives the instruction 'change state', and when the clock goes high the slave flip-flop executes the

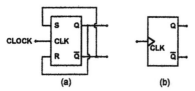

Figure 10.4 A clocked 'toggle' or 'T-type' flip-flop is constructed as shown in *(a)*, and uses the standard symbol of *(b)*

instruction, so the output changes state on the arrival of the leading edge of each new clock pulse, and two clock pulses are needed to complete a full switching cycle; the output switching frequency is thus half that of the clock frequency, and this circuit, which is known as a 'toggle' or 'T-type' flip-flop, thus acts as a binary divide-by-2 counter.

Figure 7.4(b) shows the basic circuit symbol of the clocked T-type flip-flop; note that the sharp-edged 'notch' symbol on the CLK input indicates that the flip-flop is triggered by the rising edge of a clock signal (falling-edge triggering can be notated by adding a 'little circle' symbol to the CLK line).

D and JK Flip-flops

The T-type flip-flop acts purely as a counter/divider. A far more versatile device is the 'data' or D-type flip-flop, which is made by connecting the clocked master–slave flip-flop as shown in *Figure 10.5*. Here, the inverter wired between the flip-flop's S and R terminals ensures that the signal on the DATA input is applied to these pins in anti-phase. *Figures 10.5(b)* and *10.5(c)* show the symbol and Truth Table of the D-type flip-flop, which can be used as a data latch by using the connections shown in *Figure 10.6(a)*, or as a binary counter/divider

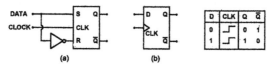

Figure 10.5 Basic circuit *(a)*, symbol *(b)*, and Truth Table of the D-type flip-flop

by using the connections shown in *Figure 10.6(b)* (with the D and NOT-Q terminals coupled together).

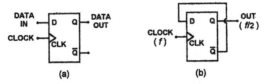

(a)   (b)

Figure 10.6   A D-type flip-flop can be used as *(a)* a data latch or *(b)* as a divide-by-2 (binary counter/divider) circuit

Figure 10.7 shows the basic circuit, symbol and action table of an even more important and versatile clocked flip-flop, which is universally known as the JK-type flip-flop. This flip-flop can be 'programmed' to act as either a data latch, a counter/divider, or as a do-nothing element by suitably connecting the J and K terminals as indicated in the table. In essence, the JK flip-flop acts like a T-type when both J and K terminals are high, or as a D-type when the J and K terminals are at different logic levels. When both J and K terminals are low the flip-flop states remain unchanged on the arrival of a clock pulse.

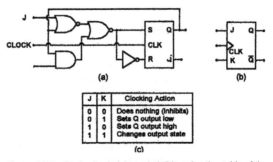

(a)   (b)

J	K	Clocking Action
0	0	Does nothing (inhibits)
0	1	Sets Q output low
1	0	Sets Q output high
1	1	Changes output state

(c)

Figure 10.7   Basic circuit *(a)*, symbol *(b)*, and action table of the JK flip-flop

Practical TTL Flip-flop ICs

One of the best known TTL flip-flop ICs is the 74LS74. This 'Dual' IC houses two independent D-type flip-flops that share common power supply connections; *Figure*

10.8 shows the IC's functional diagram and Truth
Tables. Before looking at ways of using this IC, it is
worth spending a few moments considering the general
symbology, etc., used in this and other flip-flop
diagrams, as follows.

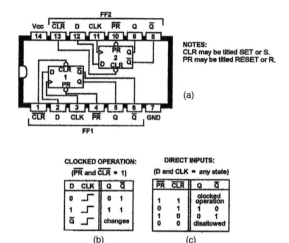

Figure 10.8 Functional diagram and Truth Tables of the 74LS74
Dual D-type flip-flop IC

Note in functional diagram *(a)* of *Figure 10.8* that each
flip-flop element has inputs that are internally notated
PR and CLR, and that the feed-ins to these points carry
little negation circles, indicating that PR and CLR are
active-low; consequently, the actual IC pins that connect
to these points should correctly be notated $\overline{PR}$ and $\overline{CLR}$
(NOT-PR and NOT-CLR) to indicate their active-low
actions. The terms PR and CLR are actually
abbreviations for PRESET and CLEAR, and are
favoured by most manufacturers of flip-flop ICs, but –
as is shown in Truth Table *(c)* – they give the same
direct control of the Q and NOT-Q output states as
conventional SET and RESET signals, and these latter
terms are preferred by a few manufacturers.

Figure 10.9 shows four common variations of the
74LS74 D-type flip-flop symbol. The *(a)* diagram states
that the flip-flop uses rising-edge clocking and negated
PR and CLR control, etc., and is an excellent 'data

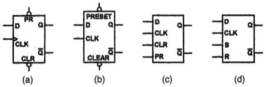

Figure 10.9   Four common variations of the 74LS74 D-type flip-flop symbol

sheet' diagram; but all D-type flip-flops are edge-triggered, and the PR and CLR notations are a bit ambiguous, so the diagram of *(b)* is equally good. The *(a)* and *(b)* symbols are both too complex for general circuit-diagram use, and can be simplified by eliminating their negation circles, etc., and abbreviatiating and perhaps repositioning their PRESET/CLEAR or SET/RESET titles, as shown in the examples of *(c)* and *(d)*.

Turning now to ways of using the 74LS74 or similar D-type flip-flop elements, note from *Figure 10.8*'s Truth Table *(c)* that $\overline{PR}$ and $\overline{CLR}$ must be tied to logic-1 to give normal clocked operation, and from Truth Table *(b)* that the element can be used as a Data latch by feeding Data to the D terminal, or as a divide-by-two counter by shorting the D and NOT-Q terminals together. *Figure 10.10* shows the 74LS74 connections used in these two modes of operation; in the Data latch *(a)* mode the 'D' logic-level is latched into the flip-flop and presented at its Q output by the rising edge of a clock pulse, and is then retained until another clock pulse arrives.

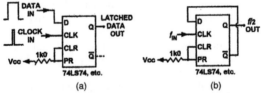

Figure 10.10   A Data latch *(a)* and a divide-by-2 counter *(b)* made from a 74LS74 or similar D-type flip-flop

Another popular TTL Dual flip-flop IC is the 74LS73, which has the functional diagram and Truth Tables shown in *Figure 10.11*. Note this IC's unusual supply-pin connections, with pin-4 acting as V_{cc} and pin-11 as

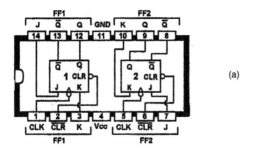

(a)

DIRECT INPUTS:
(J and K = any state)

CLR	CLK	Q	Q̄
0	0 or 1	0	1
1	1	no change	

CLOCKED OPERATION:
(CLR = 1)

J	K	CLK	Q	Q̄
0	0	⏄	no change	
1	0	⏄	1	0
0	1	⏄	0	1
1	1	⏄	changes	

(b)   (c)

Figure 10.11 Functional diagram and Truth Tables of the 74LS73 Dual JK flip-flop IC

GND. Also note that, like most TTL JK flip-flops, it uses negative-edge triggering, as indicated by the little negation circle on the clock input of each flip-flop symbol. Each flip-flop has a negated CLR (CLEAR) terminal, which is normally biased to logic-1; when CLR is pulled to logic-0 it sets the Q output to logic-0. *Figure 10.12* shows three variations of the 74LS73 JK flip-flop symbol; *(a)* is a strictly correct 'Data Sheet' type of symbol, and *(b)* and *(c)* are simplified versions suitable for use in circuit diagrams, etc.

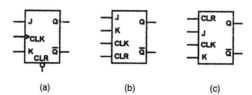

(a)   (b)   (c)

Figure 10.12 Strictly correct *(a)* and common variations *(b)* and *(c)* of the 74LS73 JK flip-flop symbol

Figure 10.13 shows two basic ways of using a 74LS73 JK flip-flop element. It can be used as a divide-by-2

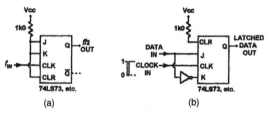

Figure 10.13 A divide-by-2 circuit *(a)* and a Data latch *(b)* made from a 74LS73 or similar JK flip-flop

circuit by tying its CLR and J and K terminals to logic-1 via a shared bias resistor and feeding the input signal to the CLK terminal, as shown in *(a)*, or as a Data latch by biasing CLR to logic-1, feeding the Data in direct form to J and inverted form to K, and using a negative-going pulse to latch the data in, as shown in *(b)* (note that the Standard '7473' version of this IC differs slightly, and is triggered by the falling edge of a positive pulse).

There are two useful variations of the basic 74LS73 IC. One of these is the 74LS107, which is internally identical to the 74LS73 but has more conventional pin allocations, with pin-7 acting as GND, and pin-14 as V_{cc}; *Figure 10.14* shows this IC's functional diagram, etc. The other variation is the 74LS76, which uses a 16-pin IC package in which each flip-flop is provided with a negated PRESET (PR) terminal as well as a CLEAR (CLR) one; *Figure 10.15* shows the functional diagram of this IC.

High on the list of other popular TTL flip-flop ICs are the 74LS175 Quad D-type, the 74LS273 Octal D-type, and the 74LS374 Octal D-type. In each of these ICs, all

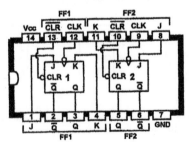

Figure 10.14 Functional diagram of the 74LS107 Dual JK flip-flop IC

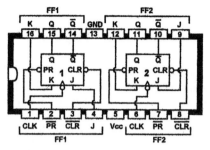

Figure 10.15 Functional diagram of the 74LS76 Dual JK flip-flop IC with PRESET and CLEAR

flip-flops are effectively connected in parallel and share common CLOCK and RESET (or CLR) inputs.

Practical 74HC-series Flip-flop ICs

The best-known '74HC'-series clocked flip-flop IC is the 74HC74 Dual D-type, which is simply a fast CMOS version of the popular 7474 or 74LS74 TTL types, and uses the functional diagram, etc., already shown in *Figure 10.8*. *Figure 10.16* shows the connections for using a 74HC74 flip-flop element as a data latch or a divide-by-2 counter; in both cases, the CLR and PR (i.e. S and R) terminals are tied directly to the IC's positive supply rail.

High on the list of other popular '74HC'-series flip-flop ICs are the 74HC175 Quad D-type, which is simply a fast CMOS version of the 74LS175 TTL IC, and the 74HC273 and 74HC374 Octal D-types, which are simply fast CMOS versions of the 74LS273 and 74LS374 TTL ICs.

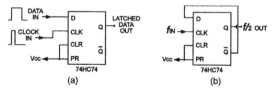

Figure 10.16 A Data latch *(a)* and a divide-by-2 counter *(b)* made from a 74HC74 (or similar) D-type flip-flop

Practical '4000'-series Flip-flop ICs

The two best-known '4000'-series CMOS clocked flip-flop ICs are the 4013B D-type and the 4027B JK-type. Both of these ICs are Duals, each containing two independent flip-flops sharing common power supply connections. *Figure 10.17* shows the functional diagram and Truth Tables of the 4013B, and *Figure 10.18* shows similar details of the 4027B.

Note that both of these ICs have SET and RESET input terminals that are additional to the connections shown in the basic *Figure 10.5* and *10.7* circuits. These terminals are known as direct inputs and enable the clock action of the flip-flop to be over-ridden, so that the devices can act as simple unclocked SET–RESET flip-flops. For normal clocked operation as a counter/divider or data latch, etc., the direct R and S terminals must be tied to logic-0, as indicated in the two

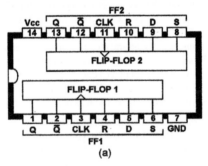

(a)

CLOCKED INPUTS

D	CLK	Q	Q̄
0	⌐⌐	0	1
1	⌐⌐	1	0
Q̄	⌐⌐	Changes	

DIRECT INPUTS

R	S	Q	Q̄
0	0	Clocked operation	
0	1	1	0
1	0	0	1
1	1	Disallowed	

(b)

Figure 10.17   Functional diagram *(a)* and Truth Tables *(b)* of the 4013B Dual D-type flip-flop IC

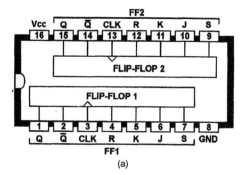

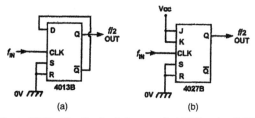

CLOCKED INPUTS					DIRECT INPUTS			
K	**J**	**CLK**	**Q**	**Q̄**	**R**	**S**	**Q**	**Q̄**
0	0	⌐⌐	No change		0	0	Clocked operation	
0	1	⌐⌐	1	0	0	1	1	0
1	0	⌐⌐	0	1	1	0	0	1
1	1	⌐⌐	Changes		1	1	Disallowed	
		(b)					(c)	

Figure 10.18 Functional diagram *(a)* and Truth Tables *(b)* of the 4027B Dual JK flip-flop IC

sets of Truth Tables, and as shown in *Figure 10.19*, which shows the flip-flops configured as divide-by-2 counter/divider elements.

The 4013B and 4027B are fast-acting ICs, and when using them it is very important to note that their clock signals must be absolutely clean (noise-free and bounceless) and have rise and fall times of less than 5µS. The 4013B is particularly fussy about the shape of

Figure 10.19 Specific circuit diagrams for '4000'-series CMOS *(a)* D-type and *(b)* JK-type divide-by-2 stages

its input clock signals, the 4027B being rather less fussy about such matters. Both devices clock or 'shift' on the positive transition of the clock signal.

Circuit Diagram Symbology

It is pertinent at this moment to note a few aspects of logic symbology, as applicable to clocked flip-flops and circuit diagrams. Broadly speaking, any logic-circuit diagram can be presented in either a generalized form, using greatly simplified circuit symbols, or in a specific form, using accurate circuit symbols that apply to specific ICs. Thus, all D-type flip-flops have D and CLK input terminals and Q and NOT-Q output terminals, and all JK types have J, K and CLK inputs and Q and NOT-Q outputs, so the basic ways of using *any* D-type or JK-type flip-flop in the divide-by-2 mode can be presented as in the general or 'universal' circuit diagram of *Figure 10.20*. If an engineer wants to build either of these circuits from a specific IC of the appropriate type he (or she) can do so by simply looking at that IC's Data Sheet or Truth Table to find the appropriate connections for any terminals that are not mentioned in the general diagram, and then adding that data to a 'specific' circuit diagram, as already shown in the examples of *Figures 10.16(b)* and *10.19*.

Thus, 'general' logic-circuit diagrams are a useful way of presenting valuable design information, and can easily be translated into 'specific' circuit diagram form. Several 'general' logic-circuit diagrams are used in the remaining sections of this chapter.

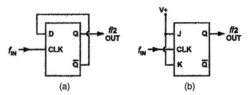

(a)   (b)

Figure 10.20   General circuit diagram of *(a)* D-type and *(b)* JK divide-by-2 stages

Ripple Counters

The most popular application of the clocked flip-flop is as a binary counter, and the reader has already seen how individual D-type and JK-type elements can be used to give divide-by-2 action in this mode; when these circuits are clocked by a fixed-frequency waveform they give a symmetrical squarewave output at half of the clock frequency.

Numbers of basic divide-by-2 stages can be cascaded to give multiple binary division by simply clocking each new stage from the appropriate output of the preceding stage. Thus, *Figure 10.21* shows (in 'general' form) how two D or JK stages can be cascaded to give an overall division ratio of four (2^2), and *Figure 10.22* shows how three stages can be cascaded to give a division ratio of eight (2^3). *Figure 10.23* shows how D-type stages can be cascaded to make a divide-by-2^N

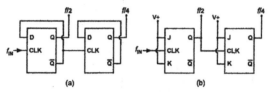

Figure 10.21   General circuit diagram of *(a)* D-type and *(b)* JK divide-by-4 ripple counters

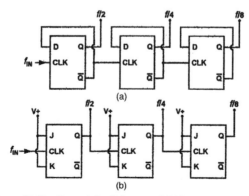

Figure 10.22   General circuit diagram of *(a)* D-type and *(b)* JK divide-by-8 ripple counters

counter, where 'N' is the number of counter stages. Thus, four stages give a ratio of 16 (2^4), five stages give 32 (2^5), six give 64 (2^6), and so on. In modern flip-flop jargon, the number of stages in a multi-stage binary divider are often referred to as its 'bit' size; thus, a four-stage counter may be called a '4-bit' counter, or an eight-stage counter an 8-bit counter, and so on.

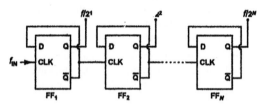

Figure 10.23   General circuit diagram of a D-type divide-by-2^N ripple counter

The multi-bit circuits in *Figures 10.21* to *10.23* are known as ripple dividers (or counters), because each stage is clocked by a preceding stage (rather than directly by the input clock signal), and the clock signal thus seems to 'ripple' through the dividers. Inevitably, the propagation delays of the individual dividers all add together to give a summed delay at the end of the chain, and divider stages (other than the first) thus do not clock in precise synchrony with the original clock signal; such divider/counters are thus asynchronous in action.

If the multi-bit outputs of a ripple divider/counter are decoded via gate networks, the propagation delays of the asynchronous dividers can result in unwanted output 'glitches' (see the 'Output decoding' section later in this chapter), so ripple counters are best used in straightforward frequency-divider applications, where no decoding is required.

Practical Ripple Counter ICs

TTL or CMOS 'Dual' flip-flop ICs can be cascaded to give any desired number of ripple stages, but where more than four stages are needed it is invariably more economic to use special-purpose MSI ripple-counter

ICs. *Figure 10.24* shows one popular but rather expensive TTL IC of this type, the 74LS93 4-bit JK ripple counter/divider, in which the four dividers are arranged in 1-bit plus 3-bit style and share a common 2-

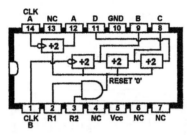

Figure 10.24 Functional diagram of the 74LS93 4-bit JK ripple counter/divider IC

input AND gated RESET line that puts all four outputs (A, B, C, D) into the '0' state when both inputs (R1 and R2) are high, but gives normal operation when one or both inputs are low. *Figure 10.25(a)* shows how to use this IC as a 3-bit ripple counter, with the A stage disabled and the B–C–D stages in use, and *(b)* shows how to use it as a 4-bit counter, with the output of the A stage feeding into the input of the B–C–D chain.

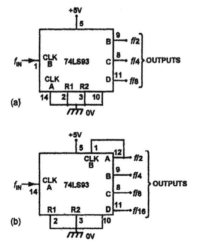

Figure 10.25 Connections for using the 74LS93 as *(a)* a 3-bit or *(b)* 4-bit ripple counter/divider

Figure 10.26 shows the functional diagram of a far more economic TTL IC, which (at the time of writing) costs half the price of the 74LS93 but has double the bit count. This is the 74LS393 Dual 4-bit ripple-counter IC, in which each 4-bit counter has its own CLOCK and RESET (RS) input pins; the RESET line puts all four outputs (A, B, C, D) into the '0' state when the RS input is high, but gives normal ripple-count operation when RS is low. *Figure 10.27(a)* shows how to use this IC as a

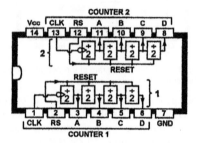

Figure 10.26   Functional diagram of the 74LS393 Dual 4-bit ripple counter/divider IC

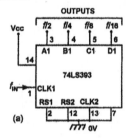

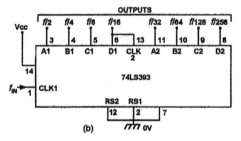

Figure 10.27   Basic ways of using the 74LS393 as *(a)* a 1- to 4-bit circuit, or *(b)* a 1- to 8-bit circuit

1- to 4-bit ripple counter, with the '1' counter in use and the '2' counter disabled, and *(b)* shows how to use it as a 1- to 8-bit counter, with both counters in use.

Figure 10.28 shows the functional diagram of one popular CMOS ripple counter IC, the 4024B or its 'HC' counterpart, the 74HC4024. This is a 7-stage ripple unit

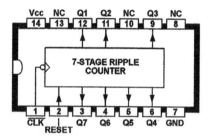

Figure 10.28 Functional diagram of the 4024B or 74HC4024 7-stage ripple counter IC

with all seven outputs externally accessible, and gives a maximum division ratio of 128. *Figure 10.29* shows another popular CMOS unit, the 4040B (or 74HC4040); this is a 12-stage unit with all outputs accessible, and

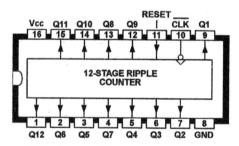

Figure 10.29 Functional diagram of the 4040B or 74HC4040 12-stage ripple counter IC

gives a maximum division ratio of 4096. *Figure 10.30* shows the functional diagram of a 14-stage CMOS unit, the 4020B (or 74HC4020), which gives a maximum division ratio of 16,384 and has all outputs except 2 and 3 externally accessible.

Finally, *Figure 10.31* gives details of another 14-stage unit, the 4060B or 74HC4060 IC. This IC does not have

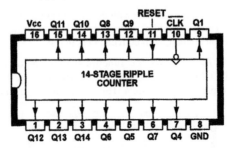

Figure 10.30 Functional diagram of the 4020B or 74HC4020 14-stage ripple counter

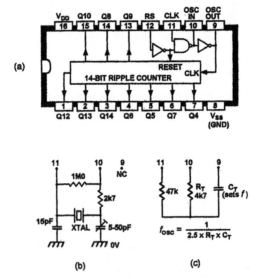

Figure 10.31 Functional diagram (a) of the 4060B or 74HC4060 14-stage ripple counter, with connections for using its internal gates as (b) a crystal oscillator or (c) R–C oscillator

outputs 1, 2, 3 and 11 externally accessible, but incorporates its own built-in clock oscillator circuit. The diagram shows the connections for using the internal circuit as either a crystal or an R–C oscillator.

Note that all eight of these CMOS ripple-counter ICs are provided with Schmitt-like trigger action on their input terminals, and can thus be clocked via relatively slow or non-rectangular input waveforms. They all trigger on the negative transition of each input cycle. All counters can be set to zero by applying a high level (logic-1) to the RESET line.

Figures 10.32 to *10.34* show three typical examples of large-bit ripple divider applications. In *Figure 10.32* a 'top-C' (4186.0Hz) generator is used in conjunction with a 7-bit ripple divider (such as the 4024B) to make an 8-octave 'C' note generator that produces symmetrical outputs on terminals C1 to C7; this basic type of circuit is widely used in electronic pianos, etc. *Figure 10.33* shows a 15-bit ripple divider and a 32.768kHz crystal oscillator used to make a precision 1Hz timing generator of the type that is commonly used in digital watches, and *Figure 10.34* uses a 22-bit ripple divider and a commonly available 4.194304MHz crystal reference, etc., to make another precision 1Hz generator.

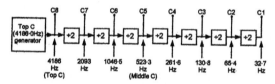

Figure 10.32   A 7-bit ripple divider used to make an 8-octave 'C'-note generator

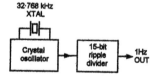

Figure 10.33   Timing generator circuit commonly used in digital watches

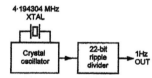

Figure 10.34   Timing generator circuit using a commonly available crystal reference

Output Decoding

The outputs of a 2-stage divide-by-4 ripple counter
have, as is shown in *Figures 10.35(a)* and *(b)*, four
possible binary states. Thus, at the start or '0'-reference
point of each clock cycle the Q2 and Q1 outputs are
both in the logic-0 state. On the arrival of the first clock
pulse in the cycle, Q1 switches high. On the arrival of
the second pulse Q2 goes high and Q1 goes low. On the
third pulse Q2 and Q1 both go high. Finally, on the
arrival of the fourth pulse Q2 and Q1 both go low
again, and the cycle is back to its original '0'-reference
state.

Each of the four possible coded states of the ripple
counter can be decoded, to give four unique outputs, by
ANDing the logic-1 outputs that are unique to each one
of the states, using the AND gate connections shown in
Figure 10.35(c). Since the ripple counter is an
asynchronous device, however, the propagation delay

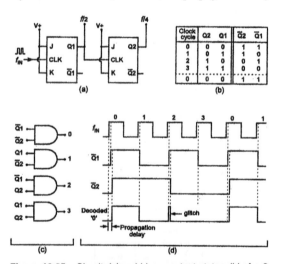

Figure 10.35   Circuit *(a)* and binary output states *(b)* of a 2-
stage ripple counter   Each of the four possible binary states can
be decoded via a 2-input AND gate *(c)*, but the decoded outputs
may not be glitch-free, as shown in the decoded-'0' example in *(d)*

between the two flip-flops may cause 'glitches' to appear in some decoded outputs, as in the example of the decoded-'0' waveforms shown in *Figure 10.35(d)*.

The principles outlined in *Figure 10.35* can be extended to any multi-stage ripple counter in which all significant binary outputs are accessible for decoding. Note, however, that the greater the number of stages, the greater become the total propagation delays and, consequently, the greater the magnitude (width) of any decoded glitches.

Walking-ring Counters

Ripple counters are very useful where simple binary division is needed, but (because of the 'glitch' problem) are not suitable for use in some sensitive decoded-counting applications. Fortunately, an alternative binary division technique is available that is well suited to the latter type of application; it is known as the 'walking-ring' technique. In this technique, numbers of flip-flops (usually JK types) are clocked in parallel and thus operate in synchrony with the input clock signal, and digital feedback determines how each stage will react to individual clock pulses; such counters are known as synchronous types, and they give glitch-free decoded outputs.

The JK version of the walking-ring technique depends on the fact that any JK flip-flop can be 'programmed' via its J and K terminals to act as either a SET or RESET latch, a binary divider, or as a do-nothing device. A detailed example of the basic walking-ring technique is given in *Figure 10.36*, which shows the circuit and Truth Tables of a synchronous divide-by-3 counter. Note that the Truth Table shows the action state of each flip-flop at each stage of the counting cycle, and remember that when the clock is low the action instruction is loaded (via the JK terminals) into the flip-flop, and the instruction is then carried out as the clock signal transitions high.

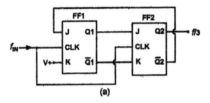

(a)

Clock cycle	Q2	Q1	FF2		FF1	
			JK code	Instruction	JK code	Instruction
⎍ 0	0	0	0 1	Set Q2 low	1 1	change state
⎍ 1	0	1				
⎍ 1	0	1	1 0	Set Q2 high	1 1	change state
⎍ 2	1	0				
⎍ 2	1	0	0 1	Set Q2 low	0 1	Set Q1 low
⎍ 0	0	0				

(b)

Figure 10.36   Circuit *(a)* and Truth Tables of a synchronous divide-by-3 counter

Thus, at the start of the cycle (CLK low), when Q2 and Q1 are both low the binary instruction 'change state' (11) is loaded into FF1 via its J and K terminals, and the instruction 'set Q2 low' (01) is loaded into FF2. On the arrival of the first clock pulse this instruction is executed, and Q1 goes high and Q2 stays low.

When the clock goes low again, new program information is fed to the flip-flops. FF1 is told to 'change state' (11), and FF2 is told to 'set Q2 high' (10), and these instructions are executed on the rising edge of the second clock pulse, causing Q2 to go high and Q1 to go low.

When the clock goes low again new program information is again fed to the flip-flops from the outputs of their partners. FF1 is instructed 'set Q1 low' (01) and FF2 is instructed 'set Q2 low' (01); these instructions are executed on the rising edge of the next clock pulse, driving Q1 and Q2 back to their original '0' states. The counting sequence then repeats *ad infinitum.*

Thus, in the walking-ring counter all flip-flops are clocked in parallel, but are cross-coupled so that the clocking response of any one stage depends on the states of the other stages. Walking-ring counters can be configured to give any desired count ratio, and *Figures 10.37* and *10.38* show the circuits and Truth Tables of divide-by-4 and divide-by-5 counters respectively. In some cases, circuit operation relies on cross-coupling

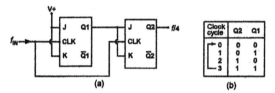

(a)

Clock cycle	Q2	Q1
0	0	0
1	0	1
2	1	0
3	1	1

(b)

Figure 10.37 Circuit *(a)* and Truth Table of a synchronous divide-by-4 counter

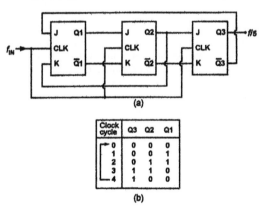

(a)

Clock cycle	Q3	Q2	Q1
0	0	0	0
1	0	0	1
2	0	1	1
3	1	1	0
4	1	0	0

(b)

Figure 10.38 Circuit *(a)* and Truth Table of a synchronous divide-by-5 counter

via AND gates, etc., and an example of this is shown in the four-stage divide-by-16 counter of *Figure 10.39*. In walking-ring counters based on D-type flip-flops, cross-coupling may be made to the SET or RESET terminals of individual flip-flop stages, etc.

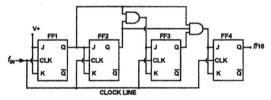

Figure 10.39 Circuit of a synchronous divide-by-16 counter

The 'Johnson' Counter

One useful variation of the synchronous walking-ring counter is a circuit known as the 'Johnson' counter, which is really a closed-loop version of a circuit known as a 'bucket brigade' synchronous data shifter. *Figure 10.40* shows a basic bucket brigade circuit; note that it is shown built from parallel-clocked cascaded D-type flip-flops, and remember that these flip-flops latch the 'D'-terminal data (logic state) into 'Q' on the arrival of the rising edge of a clock pulse. Thus, if all flip-flop outputs are initially in the logic-0 state, output A latches into the logic-1 state for one full clock cycle if a brief edge-straddling 'start' pulse is fed to FF1 at the start of that cycle, as shown. At the start of the next clock cycle the A waveform latches into FF2 and appears at output B, and in the next cycle it shifts down to C, and so on down the line for as many flip-flop stages as there are, until it is eventually clocked out of the circuit. Thus, the initial 'A' waveform passed through the circuit one step at a time, bucket-brigade style, in synchrony with the clock signal.

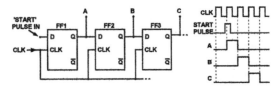

Figure 10.40   Circuit and waveforms of a basic bucket brigade synchronous data shifter

Figure 10.41 shows the basic circuit and waveforms of a 3-stage Johnson counter which is simply a bucket brigade circuit with its FF3 NOT-Q output looped back to FF1's 'D' terminal. To understand the circuit operation, assume that the Q outputs of all three flip-flops are initially set at logic-0; the D input of FF1 is thus at logic-1. On the arrival of CP1 (clock pulse 1), output A latches high, and B and C remain low. On the arrival of CP2 the NOT-Q output of FF3 is still high, so output A remains in the high state, but output B is also

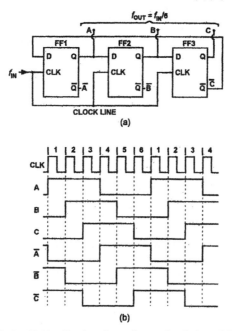

Figure 10.41 Circuit and waveforms of a 3-stage Johnson counter

latched high. On the arrival of CP3 the NOT-Q output of FF3 is still high, so outputs A and B remain high, but output C is also latched high. Consequently, on the arrival of CP4 the NOT-Q output of FF3 is low, so output A latches into the low state, but outputs B and C remain high. This process continues, as shown by the *Figure 10.41* waveforms, until the end of CP6, and a new sequencing process then starts on the arrival of the next clock pulse.

There are several important features to note about the basic Johnson counter circuit of *Figure 10.41*. First, note that a complete operating cycle takes six clock pulses (i.e. twice the number of flip-flop stages), and that the output waveform taken from any one of the six available output points (A, B, C, etc.) is a symmetrical square-wave with an operating frequency of $f_{IN}/6$. Next, note that if the A output waveform is taken as a reference point, all other output waveforms are effectively phase-shifted by 360/6 = 60 degrees relative to A. Finally, note that the outputs of the circuit can be decoded via a 2-input AND gate to give a logic-1 output for the duration of any specific clock cycle by using the connections shown in *Figure 10.42*; thus, to decode CP3, simply AND outputs A and C, or to decode CP5 simply AND C and NOT-B, and so on.

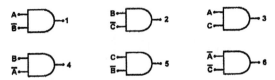

Figure 10.42   Clock-cycle decoding networks for use with the *Figure 10.41* circuit

The *Figure 10.41* circuit is shown built from three D-type flip-flop stages, but in practice a standard Johnson counter can be built from either D-type or JK flop-flops, can have any desired number of stages, and gives output frequency and phase shift magnitudes that are directly related to the total number of flip-flop stages. *Figure 10.43* shows the basic form of the two types of circuit, together with the basic formula that is common to both circuits, and *Figure 10.44* lists the basic performance

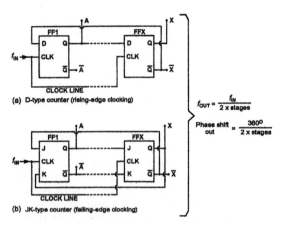

Figure 10.43   Basic D-type and JK versions of the Johnson counter, with formulae

Stages	f_{IN}/f_{OUT}	Phase shift	Comments
2	4	90°	'Quadrature' generator
3	6	60°	Popular '3-phase' (120°) generator
4	8	45°	————————
5	10	36°	Popular 'decade counter' format
6	12	30°	————————

Figure 10.44 Basic performance details of some standard Johnson counter circuits

details of various 'standard' Johnson counter circuits. Note from *Figure 10.44* that the 2-stage circuit is sometimes called a 'quadrature' generator, since it gives four outputs that are phase-shifted by 90 degrees relative to each other, and that the 3-stage circuit (see *Figure 10.41*) can be used as a 3-phase (120 degree phase shift) squarewave generator by taking outputs from A, C, and NOT-B. The 5-stage circuit is widely used as a synchronous decade counter/divider (see Chapter 11).

All the Johnson counter circuits listed in *Figure 10.44* give even values of frequency division (4, 6, 8, etc.), but such counters can also be made to give odd division values by varying their feedback connections, as has already been shown in the divide-by-3 and divide-by-5 examples of *Figures 10.36* and *10.37*. In practice, one easy way of making a circuit of this type is to use a dedicated programmable Johnson counter IC such as the 4018B.

The 4018B Divide-by-*N* counter

Figure 10.45 shows the functional diagram and outline of the 4018B 'presettable' divide-by-*N* counter IC, which can be made to divide by any whole number between 2 and 10 by merely cross-coupling its DATA and output terminals in various ways. The 4018B incorporates a 5-stage Johnson counter, has a built-in Schmitt trigger in its clock line, and clocks on the positive transition of the input signal. The counter is said to be 'presettable' because its outputs can be set to a desired state at any time by feeding the inverted version

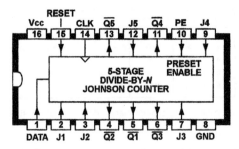

Figure 10.45 Functional diagram of the 4018B presettable divide-by-*N* counter IC

of the desired binary code to the J1 to J5 'JAM' inputs and then loading the data by taking the PRESET ENABLE (pin-10) terminal high.

Figure 10.46 shows methods of connecting the 4018B to give any whole-number division ratio between 2 and 10. On even division ratios no additional components are needed, but on odd ratios a 2-input AND gate is needed in the feedback network.

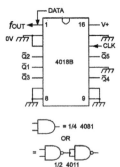

DIVISION RATIO	FEEDBACK CONNECTIONS
2	$\overline{Q}1$ to DATA
3	$\overline{Q}1$–, $\overline{Q}2$– ⟩ to DATA
4	$\overline{Q}2$ to DATA
5	$\overline{Q}2$–, $\overline{Q}3$– ⟩ to DATA
6	$\overline{Q}3$ to DATA
7	$\overline{Q}3$–, $\overline{Q}4$– ⟩ to DATA
8	$\overline{Q}4$ to DATA
9	$\overline{Q}4$–, $\overline{Q}5$– ⟩ to DATA
10	$\overline{Q}5$ to DATA

Figure 10.46 Methods of connecting the 4018B for divide-by-2 to divide-by-10 operation

Greater-than-10 Division

Even division ratios greater than ten can usually be obtained by simply ripple-wiring suitably scaled standard counter stages, as shown in the examples of

Figure 10.47. Thus, a divide-by-2 and a divide-by-6 stage give a ratio of 12, a divide-by-6 and divide-by-6 give a ratio of 36, and so on. Of the examples shown, the

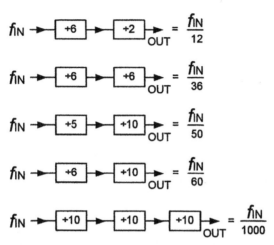

Figure 10.47 Typical examples of division by numbers greater than ten.

divide-by-50 and divide-by-60 counters are of particular value in converting 50Hz or 60Hz power-line frequencies into 1Hz timing signals with excellent long-term accuracy, and the multi-decade counters are of great value in generating precise decade-related signal frequencies from a single master oscillator.

Non-standard and uneven division ratios can be obtained by using standard synchronous counters (such as the 4018B) and decoding their outputs to generate suitable counter-reset pulses on completion of the desired count. More advanced types of counter, together with special decoder ICs, are described in the next two chapters.

11 Binary and Decade Counter ICs

The last chapter took a detailed look at D-type and JK flip-flops and showed how they can be used to make various simple counting/dividing circuits. The present chapter continues this theme by looking at some practical examples of modern TTL and CMOS binary and decade counter/divider ICs. The chapter starts off, however, by explaining some more counter/divider basics.

'Count' Direction

The design of all the multi-stage clocked flip-flop counter/dividers described in Chapter 10 is such that, in each counting sequence, each of them starts its count with all binary outputs set to the '0' state, and the outputs then increment upwards on successive clock pulses until the maximum count is reached, and the whole counting process then restarts on the arrival of the next pulse. Circuits of this type are known as 'up-counters'; all ICs described in the present chapter are up-counting types. Counter/dividers can, however, easily be designed to give a down-counting action, and a number of popular 'down' and 'up/down' counter/divider ICs are described in Chapter 12.

Synchronous or Asynchronous Counting?

In an asynchronous ripple counter, the output of one counting stage provides the clocking signal for the next stage, and as a result of each stage's propagation delay this action may cause unwanted glitches to appear on decoded outputs *as the counters switch between one set of states and another.* This glitching problem does not occur with synchronous counters, which are all clocked by the same input signal; in theory, therefore, it seems that synchronous counters are technically superior to asynchronous types. In practice, however, 'glitching' is a purely *transitory* occurrence, and the only time it is of real importance is when a decoded output is used to *directly* drive some sort of clocked logic network, and this situation occurs in relatively few practical applications.

Thus, in most real-world applications, the synchronous counter has no practical advantage over the asynchronous type, and the design engineer should – if he/she is sure that a synchronous type is not vital for use in a particular application – simply select the most cost-effective counter IC that can be used in the circuit in question. In many cases this will be one of the several decade or binary counter/divider ICs described in this chapter. Before looking at these ICs, however, note the following points about counter 'coding' systems.

Counter 'Coding' Systems

A flip-flop counter has only two possible output states (either logic-0 or logic-1), and this number can be expressed as 2^N, where N is the number of flip-flop stages and in this case equals one. This simple formula holds true for any number of cascaded flip-flop counter stages; thus, if four stages are cascaded as shown in *Figure 11.1(a)*, in which the circuit's four outputs are notated A, B, C and D, the *maximum possible* number of different ABCD 'binary code' output states equals $2^4 = 16$. Note that this number of output states is only available if the counters are wired in the ripple mode; if they are connected in the Johnson mode, only eight different possible output states are available.

Note in the *Figure 11.1(a)* circuit that counter A changes state on the arrival of each new clock (f_{IN}) pulse, and that counter D changes state on the arrival of every eighth clock pulse. Consequentially, when the ABCD

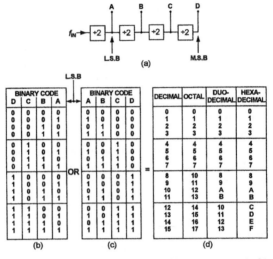

Figure 11.1   A 4-bit binary code *(a)* can be represented as in *(b)* or *(c)*, and can be translated into any of the code systems shown in *(d)*

code is used to represent a 4-bit binary number (such as '1000', etc.) it is vital to remember that 'A' represents the Least Significant Bit (LSB) of that number, and 'D' represents its Most Significant Bit (MSB). In practice, a table of multi-bit binary codes may be written with the LSB either on the right, as in the DCBA format shown in *(b)*, or on the left, as in the ABCD format shown in *(c)*; neither system has any real advantage over the other, so long as it is always remembered that 'A' represents the code's LSB. Similarly, when writing a binary code number such as '1110' in a report, etc., it is vital to make it clear to the reader whether this is an ABCD or a DCBA code, since '1110' may represent the 8th clock pulse in ABCD format or the 15th pulse in DCBA format, etc.

The tables of *Figures 11.1(b)* and *(c)* list all 16 possible 4-bit binary codes, in both DCBA and ABCD format. Each of these codes can be used, via suitable decoding/activation circuitry, to represent or perform any action or thing that you care to think of (i.e. sound an alarm, unlatch a door, display the number '7', etc.). In practice, however, they are usually used to represent numbers or letters in one or other of the four standard code systems listed in table *(d)*, in which the decimal system has a base of ten, the octal has a base of eight, the duodecimal a base of 12, and the hexadecimal (usually called 'hex') a base of 16.

The relevance of all this is that many popular counter ICs are internally configured to give outputs that comply with one or other of these code systems, and are supported by a variety of decoding/translating ICs. The following section describes some popular counter configurations.

Counter Configurations

Most popular counter/divider ICs are designed to give either decade (divide-by-10) division, or some type of binary division. In the former case, the counter can be internally configured in a variety of basic ways. *Figure 11.2*, for example, shows (in very basic form) a synchronous decade counter made from a 5-stage Johnson counter; note that this type of counter gives a symmetrical 'carry' output, and that the IC incorporates decoder circuitry that provides some sort of 'decoded' output. The most popular IC of this type is the 4017, which provides ten fully decoded outputs; this IC is described in detail later in this chapter.

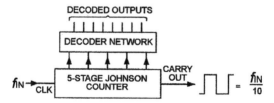

Figure 11.2 Synchronous decade counter based on a 5-stage Johnson counter

A decade counter can also take the form of a 4-stage synchronous or asynchronous counter and decoder network, configured in the basic way shown in *Figure 11.3*, in which the 4-bit decoded output conforms to the

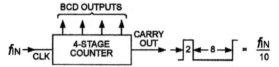

Figure 11.3 Basic 4-stage decade counter with BCD outputs

popular BCD (binary coded decimal) format, which can be used to drive a 7-segment LED or LCD display via a BCD-to-7-segment decoder/driver IC, or up to ten independent devices via a BCD-to-decimal decoder IC, etc. This type of counter generates an asymmetrical CARRY output with a 2:8 Mark–Space ratio.

The most popular type of decade counter configuration is that shown in *Figure 11.4*, in which decade division is obtained via a divide-by-2 and a divide-by-5 counter wired in series. This 'bi-quinary' configuration is quite versatile; the counter can be configured to generate a 4-bit BCD output and an asymmetrical 'carry' output by wiring it as shown in *Figure 11.4(a)*, or as a decade divider with a symmetrical 'carry' output by wiring it as shown in *Figure 11.4(b)*. One of the best known TTL ICs of this type is the 74LS90 decade divider, which is described in detail in the next section of this chapter.

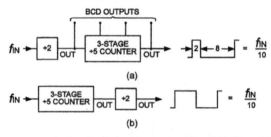

Figure 11.4 Bi-quinary decade counter giving *(a)* BCD or *(b)* symmetrical outputs

The 74LS90 IC

The 74LS90 is an asynchronous programmable counter/divider IC that contains independent divide-by-

2 and divide-by-5 counters that are triggered by clock pulse falling edges and can be used together to give decade (divide-by-10) counting with or without BCD outputs, or can be configured to give any whole-number division value from 2 to 9 inclusive. *Figure 11.5* shows the IC's functional diagram and its R/S (RESET/SET) Truth Table. Note that the two counters share SET and RESET lines that are controlled via 2-input AND gates; normally, each of these AND gates is disabled by tying at least one terminal low; the gates are active only when both inputs (R1–R2 for RESET, S1–S2 for SET) are driven high; when RESET is active, the DCBA outputs are driven to '0000' (= BCD '0'); when SET is active, the DCBA are driven to '1001' (= BCD '9'); the SET control has priority over RESET.

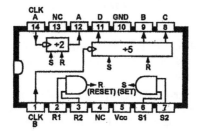

R/S INPUTS				OUTPUTS				
R1	R2	S1	S2	D	C	B	A	
1	1	0	X	0	0	0	0	= BCD '0'
1	1	X	0	0	0	0	0	
X	X	1	1	1	0	0	1	= BCD '9'
0	X	0	X					
X	0	X	0		COUNTING			
0	X	X	0					
X	0	0	X					

X = DON'T CARE

Figure 11.5   Functional diagram and R/S Truth Table of the 74LS90 decade counter IC with BCD outputs

Figures 11.6 and *11.7* show two very basic ways of using the 74LS90's counting ability. In *Figure 11.6* only the internal divide-by-2 counter is used; the divide-by-5 counter and SET/RESET gates are disabled, and the IC thus functions as a simple divide-by-2 (binary) counter. In *Figure 11.7* only the internal divide-by-5 counter is

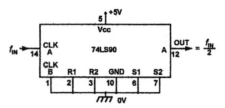

Figure 11.6   The 74LS90 used as a divide-by-2 counter

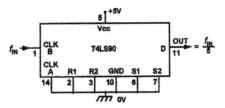

Figure 11.7   The 74LS90 used as a divide-by-5 counter

used; the divide-by-2 counter and SET/RESET gates are used, and the IC thus acts as a simple divide-by-5 (quinary) counter.

Figure 11.8 shows the IC used as a decade counter with BCD outputs. Here, both internal counters are used; f_{IN} is fed to the input of the divide-by-2 counter, and the output (A) of this drives the input of the divide-by-5 counter; in this configuration the IC functions as a divide-by-10 counter with BCD outputs. Note that the final (pin-11) output waveform is asymmetrical, with a 2:8 Mark–Space ratio.

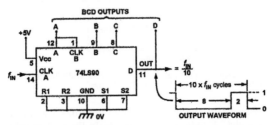

Figure 11.8   The 74LS90 used as a decade counter/divider with BCD outputs

Figure 11.9 shows the IC used as a decade counter with a symmetrical (squarewave) output. Here, both internal

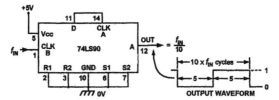

Figure 11.9 The 74LS90 used as a decade divider with a symmetrical (squarewave) output waveform

counters are again used, but f_{IN} is fed to the input of the divide-by-5 counter, and the output (D) of this drives the input of the divide-by-2 counter, which provides the final output; in this 'bi-quinary' configuration the IC thus functions as a divide-by-10 counter with a perfectly symmetrical (1:1 Mark–Space ratio) output; the outputs of this circuit are not BCD coded.

The 74LS90 can be made to divide by any whole-number value from 2 to 9 inclusive by feeding appropriate outputs back to the IC's RESET or SET line so that the counter automatically resets each time the desired count number is reached. For divide values of 3, 4, 6 and 8 these feedback connections can be taken directly from the divide-by-5 counter's DCB outputs; *Figure 11.10* shows this counter's Truth Table. Thus, the IC can be configured as a divide-

COUNT	OUTPUT		
	D	C	B
0	0	0	0
1	0	0	1
2	0	1	0
3	0	1	1
4	1	0	0

Figure 11.10 Truth Table of the 74LS90's divide-by-5 (quinary) counter

by-3 counter by using only its divide-by-5 counter, with RESET action provided via the B and C output so that it resets to '000' in the arrival of every count-3 pulse, and with the output taken from the C terminal as shown in *Figure 11.11(a)*. The IC can be made to give divide-by-6 action by using these same basic connections, but with the input applied via the divide-by-2 counter stage, as in *Figure 11.11(b)*. These two circuits can be modified to give divide-by-4 or divide-by-8 action by tying the R1

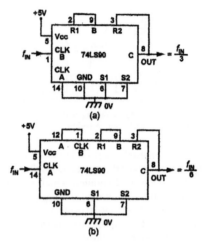

Figure 11.11 The 74LS90 used as *(a)* a divide-by-3 or *(b)* divide-by-6 counter

input to logic-1 and taking the R2 input to output D, as shown in *Figures 11.12(a)* and *(b)*.

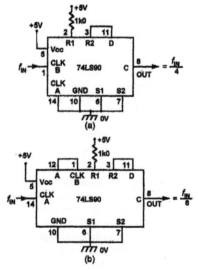

Figure 11.12 The 74LS90 used as *(a)* a divide-by-4 or *(b)* divide-by-8 counter

For divide-by values of 7 and 9, both internal counters must be used, with the divide-by-2 counter driving the divide-by-5 counter; *Figures 11.13* and *11.14* show the practical IC connections. For divide-by-7 action, both RESET terminals are grounded, but outputs B and C are coupled to the SET line via the S1 and S2 terminals, so that both counters set to the BCD-'9' state on the arrival of the 6th clock pulse, and then reset to zero when the next (7th) clock pulse arrives. For divide-by-9 action, the counters are RESET via the A and D outputs, which both go high on the 8 + 1 count.

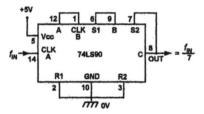

Figure 11.13   The 74LS90 used as a divide-by-7 counter

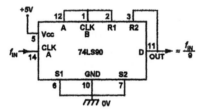

Figure 11.14   The 74LS90 used as a divide-by-9 counter

Figure 11.15 shows a practical example of a 74LS90 application circuit, in which IC1 is used to make a precision buffered-output 1MHz crystal oscillator that has its output divided down by six ripple-connected 74LS90 decade divider stages, to generate precision output frequencies of 1MHz, 100kHz, 10kHz, 1kHz, 100Hz, 10Hz, and 1Hz, which can be used as frequency or timing calibration standards, etc. Each of these 74LS90 ICs is connected in the *Figure 11.9* decade divider mode and generates a symmetrical squarewave output that is very rich in odd harmonic frequencies, i.e. the 10kHz output (for example) also

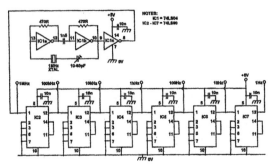

Figure 11.15   TTL six-decade crystal frequency calibrator

generates strong 30kHz, 50kHz, 70kHz, etc., output signals.

The 74LS390 IC

The 74LS390 (and its CMOS counterpart, the 74HC390) is a dual decade divider IC that can be regarded as a dual version of the basic 74LS90, but with no SET control and with greatly simplified RESET circuitry. It is cheaper than the 74LS90, and is thus a very attractive alternative to the 74LS90 in most applications where a SET facility is not needed. *Figure 11.16* shows the 74LS/HC390's functional diagram; each 'decade' counter is a bi-quinary type and consists of divide-by-2 and divide-by-5 counters that can be used independently but share a common RESET facility (RESET is normally grounded, and resets the

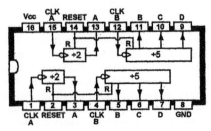

Figure 11.16   Functional diagram of the 74LS390 or 74HC390 Dual decade (bi-quinary) counter IC

decade counter to the '0000' BCDA state when taken to logic-1); the divide-by-5 counter has the same Truth Table as that used in the 74LS90 (see *Figure 11.10*).

Note in *Figure 11.16* that one decade counter appears in the lower part of the diagram, between the 'lower' (1 to 7) set of pin numbers, and the other is in the upper part, between the 'upper' (9 to 15) pin numbers; for convenience, these counters are therefore referred to in the following text by terms such as 'upper-2', 'lower-5', and 'upper-10', etc.

The 74LS/HC390 is a very easy IC to use. In each decade, the '2' and '5' counters can be used in any combination, making divide-by values of 2, 5, and 10 directly available (unwanted counters are disabled by simply grounding their CLK terminals), and each decade counter can be configured to give either a BCD or a symmetrical output. If desired, each 'decade' counter can alternatively be configured to give counts of 4 or 8 by connecting its RESET terminal to the C or D outputs as shown in *Figure 11.17*, or can be made to give counts of 3, 6, 7 or 9 by feeding various outputs back to the RESET terminal via an external 2-input AND gate as shown in *Figure 11.18*.

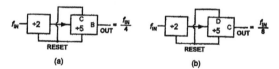

Figure 11.17 Basic ways of using a 74LS/HC390 decade divider section to give *(a)* divide-by-4 or *(b)* divide-by-8 action

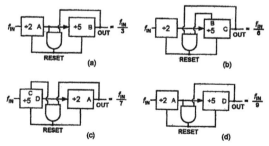

Figure 11.18 Ways of using the 74LS/HC390's decade dividers and an external 2-input AND gate to give divide-by values of *(a)* 3, *(b)* 6, *(c)* 7, or *(d)* 9

A very large total number of divide-by combinations are thus available from the 74LS/HC390 IC; *Figure 11.19* shows how values of 2, 4, 5, 10, 20, 25, 50, and 100 can be obtained by just ripple-wiring the ICs

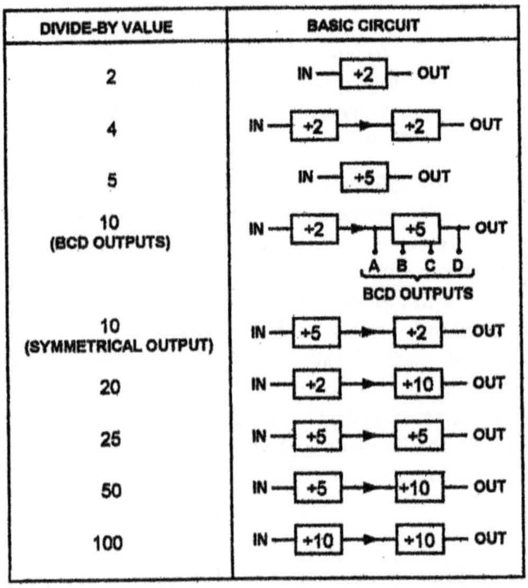

DIVIDE-BY VALUE
BASIC CIRCUIT
2
IN ÷2 OUT
4
IN ÷2 ÷2 OUT
5
IN ÷5 OUT
10 (BCD OUTPUTS)
IN ÷2 ÷5 OUT
A B C D
BCD OUTPUTS
10 (SYMMETRICAL OUTPUT)
IN ÷5 ÷2 OUT
20
IN ÷2 ÷10 OUT
25
IN ÷5 ÷5 OUT
50
IN ÷5 ÷10 OUT
100
IN ÷10 ÷10 OUT

Figure 11.19 Basic ways of using the 74LS/HC390 to give various 'divide-by' values

standard '2' and '5' counters. It would be tedious to draw up a full set of wiring diagrams for all of these divide-by combinations, but *Figure 11.20* shows how a divide-by-50 counter can be made by ripple-wiring the upper-5, lower-2, and lower-5 counters and disabling upper-2, and *Figure 11.21* shows a divide-by-100

+5V
9
1
3
4
D
CLK A
A
CLK B
Vcc
16
CLK B
12
74LS390
D
7
OUT
$= \frac{f_{IN}}{50}$
f_{IN}
GND
RESET
CLK A
RESET
8
14
15
2
0V

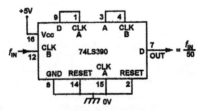

Figure 11.20 The 74LS390 used as a divide-by-50 counter

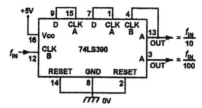

Figure 11.21 The 74LS390 used as a divide-by-100 counter with symmetrical divide-by-10 and divide-by-100 outputs

counter with symmetrical outputs made by ripple-wiring the upper-5, upper-2, lower-5, and lower-2 counters. In practice this versatile IC can, by using ANDed RESET control where necessary, be configured to give any desired whole-number divide-by value in the range 2 to 100 inclusive.

The 74HC393 Dual Binary Divider IC

The 74HC393 is a fast CMOS version of the 74LS393 Dual counter IC, and is regarded as a 'binary' brother of the 74HC390. *Figure 11.22* shows the functional diagram of the 74HC393, which is actually a Dual 4-bit ripple counter in which each 4-bit counter has its own CLK and RESET (RS) input pins and triggers on the falling edge of its CLK signal; the RESET line sets all four outputs (A, B, C, D) into the '0' state when the RS input is high, but gives normal ripple-count operation when RS is low. Thus, each counter can be used to give direct division by

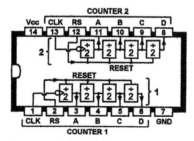

Figure 11.22 Functional diagram of the 74HC393 Dual 4-bit ripple counter/divider IC

2, 4, 8 or 16, or can be used to divide by any other number in the range 3 to 15 by suitably ANDing various outputs and feeding them back to the RESET line in a manner similar to that already shown in *Figure 11.18*.

The 74HC393's two sets of counters can be used either independently, or can be wired in series to make an 8-bit counter that can give direct division by 2, 4, 8, 16, 32, 64, 128, or 256. *Figure 11.23(a)* shows how to use the IC as a 1- to 4-bit ripple counter, with the '1' counter in use and the '2' counter disabled, and *(b)* shows how to use it as a 1- to 8-bit counter, with both counters in use.

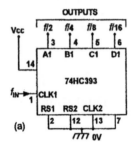

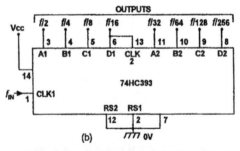

Figure 11.23 Basic ways of using the 74HC393 as *(a)* a 1- to 4-bit circuit, or *(b)* a 1- to 8-bit circuit

The 4017 Decade Counter IC

The 4017B and its '74HC'-series counterpart, the 74HC4017, is probably the most popular and useful of all

CMOS decade counter/divider ICs. It is actually a 5-stage Johnson decade counter that has ten fully decoded outputs that sequentially switch high on the arrival of each new clock pulse, only one output being high at any moment; each output can provide several milliamps of drive current to an external load. *Figure 11.24* shows the functional diagram, that applies to both versions of the IC, which has clock, reset (RS), and inhibit (INH) input terminals and ten decoded and one carry (CO) output terminals.

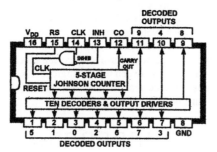

Figure 11.24 Functional diagram of the 74HC4017 Johnson decade counter with ten decoded outputs

The main differences between the two versions of the IC are that the 4017B can use any supply in the 3V to 15V range and can operate at a maximum frequency of about 12MHz when powered from a 10V supply, while the 74HC4017 is designed to use a nominal 5V supply (actually 2V to 6V) and can operate at up to 50MHz. Bearing in mind these basic differences, note that the following text and diagrams, apply specifically to the 74HC4017 version of the device.

In normal use, the 74HC4017 is connected in the basic 'decade divider' mode shown in *Figure 11.25*, with its

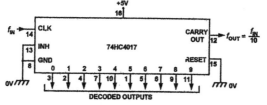

Figure 11.25 Basic connections for using the 74HC4017 as a rising-edge triggered synchronous decade divider

RESET and INH terminals grounded. In this mode the IC's Johnson counter stages advance one step on the arrival of each new clock pulse rising edge and simultaneously set one of the ten decoded outputs high while the other nine outputs remain low; the outputs go high sequentially, in phase with the clock signal, with the selected output remaining high for one full clock cycle, as shown in the waveform diagram of *Figure 11.26*. An additional carry out (CO) signal completes one cycle for every ten clock input cycles, and can be used to ripple-clock additional 74HC4017s in multi-decade counters. Note that this IC has buffer-style inputs that are not of the Schmitt type, and the IC is thus sensitive to clock waveform shapes; the clock pulses must switch fully between normal logic levels, rise and fall times must be less than 400ns, and clock pulse widths must be greater than 15ns.

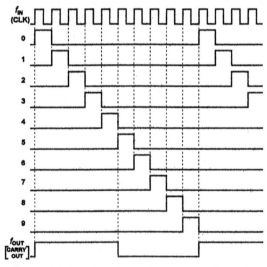

Figure 11.26 Waveform timing diagram of the *Figure 11.25* circuit

Figures 11.27 and *11.28* show two useful variations of the basic 'decade divider' circuit. *Figure 11.27* shows how to use the IC's pin-15 RESET control; normally this control is tied to logic-0, but when it is taken to

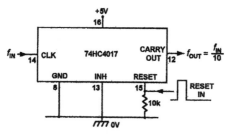

Figure 11.27   Basic way of using the 74HC4017's RESET terminal

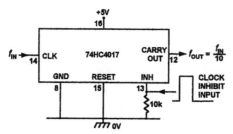

Figure 11.28   Basic way of using the 74HC4017's CLOCK INHIBIT terminal

logic-1 it resets all of the IC's counters and sets all decoded outputs except output '0' to the logic-0 state. *Figure 11.28* shows the basic way of using the pin-13 clock inhibit (INH) terminal; normally, this terminal is tied to logic-0, but when it is taken to logic-1 it fully inhibits the IC's clocking and counting actions.

There are sometimes minor architectural differences between different manufacturers' versions of the 74HC4017, and in the Philips/Mullard version these enable the device to be triggered by the rising edges of clock signals by using the connections shown in *Figure 11.25*, or by falling edges by using the connections of *Figure 11.29*, in which the CLK (pin-14) terminal is tied to logic-1 and the clock signal is fed to the INH (pin-13) terminal (in practice, this configuration works well with most versions of the 74HC4017).

Figures 11.30 and *11.31* show ways of using the 74HC4017 as a divide-by-N counter with N decoded outputs, in which N = any whole number from 2 to 9. In

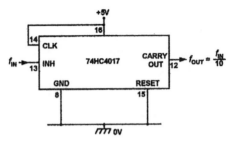

Figure 11.29   Basic way of using the Philips 74HC4017 to give falling-edge clock triggering (see text)

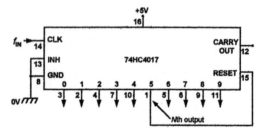

Figure 11.30   Simple way of using the 74HC4017 as a divide-by-*N* (2 to 9) counter; circuit is shown set for divide-by-5 operation

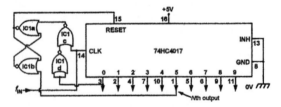

Figure 11.31   Alternative divide-by-*N* circuit, set for divide-by-5 operation

the *Figure 11.30* circuit the *N*th decoded output is simply shorted to the reset terminal so that the counter resets to zero on the arrival of the *N*th clock pulse. This circuit is slightly sensitive to the clock signal's pulse width and rise time; the *Figure 11.31* version of the counter does not suffer from this problem. Here, logic gates control the reset operation via the IC1a–IC1b flip-flop, and the action is such that the reset command is given on the arrival of the *N*th clock pulse and is maintained while the clock pulse remains high, but is removed automatically when the clock pulse goes low again. Note in both diagrams that the circuit is shown set for divide-by-5 operation; also note that the IC's 'carry out' terminal is effectively disabled when N values of less than 5 are used ('carry' signals can, however, easily be derived from the decoded '0' or N outputs).

One of the 74HC4017's most important features is its provision of ten fully decoded outputs, making the IC ideal for use in a whole range of 'sequencing' operations in which the outputs are used to drive LED displays, relays, or sound generators, etc. *Figure 11.32* shows how it can be connected to give 'sequence-and-stop' operation, in which the IC stops clocking after completing a predetermined counting sequence. In the diagram the counter is set to stop when its INH terminal is driven high by the '9' output, but it can in fact be inhibited via any one of the IC's decoded output terminals. The count sequence can be restarted by pressing reset button S1, or by feeding a positive pulse to the RESET pin.

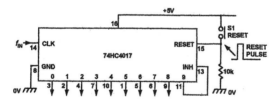

Figure 11.32   74HC4017 set for sequence-and-stop operation

The 74HC4017 can be used as a sequencing LED display driver, operating in the 'moving dot' or 'chaser' mode (in which only one LED is illuminated at any given moment), by using the basic connections shown in *Figure 11.33*, in which the IC is shown used in the divide-by-10 mode; note here that the LEDs share a common current-limiting resistor, which with the value shown limits their ON currents to about 10mA.

In applications where far more than ten LEDs need to be driven in the sequential moving-dot mode, one option is to use two 74HC4017 ICs in the basic way shown in the 'multiplexed' circuit of *Figure 11.34*,

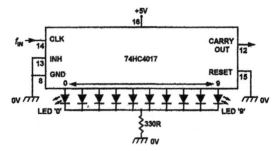

Figure 11.33 Basic way of using the 74HC4017 as a moving-dot 10-LED 'chaser' circuit

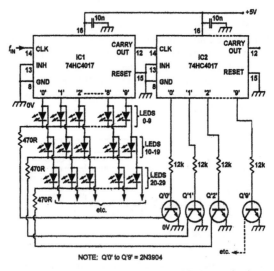

Figure 11.34 Multiplexed moving-dot LED chaser circuit can drive up to 100 LEDs

which can sequentially step-drive as many as one hundred LEDs. Here, the divide-by-10 CARRY OUT signal of IC1 is fed to the CLK input of IC2, and the LEDs are connected in banks of ten with their anodes driven by IC1 and their cathodes taken to ground via switching transistors that are activated by the outputs of IC2. The circuit's basic action is such that, for any given LED to be driven on, the appropriate outputs of both IC1 and IC2 must be high; LED12, for example, turns on only when outputs '2' of IC1 and '1' of IC2 are high, and so on. Thus, LEDs 0 to 9 are activated sequentially during clock pulses 0 to 9, and LEDs 10 to 19 are activated during pulses 10 to 19, and so on, right up to clock pulse 99, after which the whole sequence starts to repeat.

If desired, the *Figure 11.34* circuit can be made to sequence repeatedly after a count of N (rather than '99') by ANDing the appropriate outputs of IC1 and IC2 and feeding the ANDed signal directly to each IC's RESET (pin-15) terminal, which must have its ground connection removed. Thus, for an N value of 27, outputs '2' of IC2 and '7' of IC1 must be ANDed, and the circuit needs only 27 LEDs and three switching transistors. Note that the 470R resistor connected in series with the collector of each transistor acts as an LED current limiter, and can be reduced to 220R if preferred.

Finally, note that the 74HC4017 is sufficiently inexpensive to justify its use as a simple decade divider, in which case it should be connected as shown in *Figure 11.35*, with the RESET and INHIBIT terminals grounded and the output taken from the CARRY OUT terminal. *Figure 11.36*

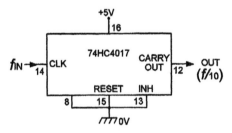

Figure 11.35 74HC4017 connected as a decade counter/divider

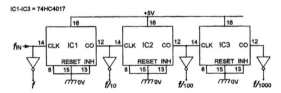

Figure 11.36 A ripple-wired 3-decade divider with buffered outputs

shows how three decade dividers of this type can be ripple-wired (cascaded) to make a 3-decade divider that generates outputs at 1/10th, 1/100th and 1/1000th of the clock frequency; the CARRY OUT signal of each counter provides the CLOCK signal to the following stage. Note in this particular circuit that the four outputs are buffered via simple CMOS inverters to ensure that output loading does not degrade the clock signal rise times. Thus, if the clock input signal is derived from a 1MHz crystal oscillator, the circuit can be used as a laboratory frequency standard, generating frequencies of 1MHz, 100kHz, 10kHz and 1kHz.

The 4022B Octal Counter IC

The 4022B can be regarded as an octal (divide-by-8) brother of the '4017'. It is a synchronous 4-stage Johnson counter with eight fully decoded outputs, so arranged that they go high sequentially. *Figure 11.37*

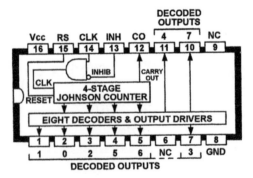

Figure 11.37   Functional diagram of the 4022B octal counter with eight decoded outputs

shows the IC's functional diagram. For normal octal counting, the RESET and INHIBIT terminals are tied low; the CARRY OUT signal completes one cycle for every eight input clock cycles.

Synchronous 'UP' Counter ICs

Synchronous 'UP' counters are normally used in fairly simple applications in which it is necessary merely to count a number of input pulses, or divide them by a fixed ratio, and (perhaps) then display the results on a 7-segment LED or LCD readout unit.

The 4017 and 4022 types of IC are popular examples of synchronous CMOS counter ICs; two other basic families of this type are also readily available, however. The oldest of these families consists of the 4026B and 4033B ICs, which are decade counters with built-in decoders that give 7-segment display outputs that can directly drive sensitive LED displays. The second family of devices are the 4518B and 4520B 'Duals', which each house two counters in a 16-pin package; the 4518B is a Dual decade counter with BCD outputs, and the 4520B is a Dual hexadecimal (divide-by-16) counter with a 4-bit binary output; the outputs of these two ICs must be decoded externally if they are to be used to drive 7-segment displays, etc. Both of these IC families are described in detail in the final two sections of this chapter.

The 4026B and 4033B Counter ICs

The 4026B and the 4033B are synchronous decade 'UP' counters that incorporate decoding circuitry that gives a 7-segment output suitable for directly driving a sensitive 7-segment common-cathode LED display; the output drive currents are limited to only a few mA. Both ICs have CLOCK, CLOCK INHIBIT and RESET input terminals, and a CARRY OUT output terminal that completes one cycle for every ten input clock cycles and can be used to clock following decades in a counting chain. The counters are advanced on the positive transition of the input clock cycle.

Figure 11.38 shows the outline and functional diagram of the 4026B. This IC features DISPLAY ENABLE input and output terminals, which enable the entire display to be blanked. Normally, the DISPLAY ENABLE IN pin is held high; the display blanks (but the IC continues to count) when this pin is pulled low. The IC also has an UNGATED 'C' SEGMENT output terminal (pin-14), which can be used with external logic to make the IC count by numbers other than ten.

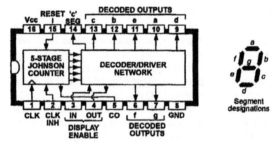

Figure 11.38 Functional diagram of the 4026B decade counter with 7-segment display driver and display enable control

Figure 11.39 shows the outline and functional diagram of the 4033B. This IC features RIPPLE BLANKING input and output terminals, which can be used to automatically blank leading and trailing zeros in multidecade applications; the '0' display blanks automatically when the RIPPLE BLANKING INPUT pin is held low. The IC also features a LAMP TEST pin, which is normally held low but which drives all seven decoded outputs high when the input pin is taken high.

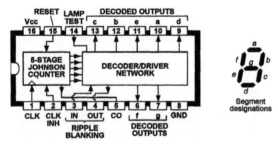

Figure 11.39 Functional diagram of the 4033B decade counter with 7-segment display driver and ripple-blanking facility

Figure 11.40 shows how to connect the 4026B and the 4033B for simple decade counter/display operation. In both cases the RESET and CLOCK INHIBIT terminals are tied low. In the case of the 4026B, the DISPLAY ENABLE INPUT must be tied high if the display is to be illuminated. In the case of the 4033B, the RIPPLE BLANKING INPUT terminal must be tied high if the display is required to give normal operation, or must be tied low if it is required to give 'zero' suppression. Note in both circuits that, if multi-decade counting is to be used, the CARRY OUT of one stage must be used to provide the CLOCK input signal of the next stage.

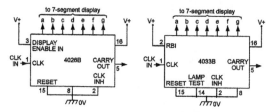

Figure 11.40 Method of connecting the 4026B and 4033B for normal decade dividing/display operation. In the case of the 4033B, the RBI connection does not give suppression of display zeros

Figure 11.41 shows how to interconnect several 4033Bs to give automatic suppression of leading and trailing zeros so that, for example, the count 009.90 is actually displayed as 9.9. To get leading zero suppression (on the integer side) the RBI (pin-3) terminal of the most significant digit (MSD) counter must be tied low, and its RBO (pin-4) terminal must be taken to the RBI terminal of the next MSD counter, and so on down to the UNITS counter. To get trailing zero suppression (on the fraction side of the display) the RBI of the least significant digit (LSD) must be tied low, and its RBO terminal must be taken to the RBI terminal of the next LSD counter, and so on, to the first counter in the 'fractions' chain.

When contemplating use of the 4026B or the 4033B, note that these ICs do *not* incorporate DATA latches; consequently, the displays tend to 'blur' when the ICs are actually going through a counting cycle. Also note that the ICs have very limited output drive current capability, and produce fairly dim LED displays; if

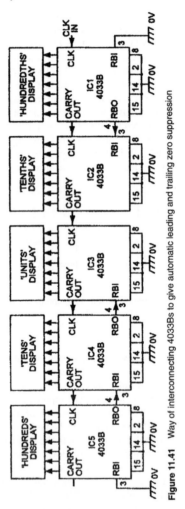

Figure 11.41   Way of interconnecting 4033Bs to give automatic leading and trailing zero suppression

desired, the LED display brightness can be greatly increased by wiring a current-boosting transistor in series with each segment-driving output terminal, as shown in *Figure 11.42*.

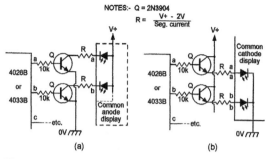

Figure 11.42 Way of boosting the current drive to *(a)* common anode or *(b)* common cathode LED displays

The 4518B and 4520B Counter ICs

The 4518B and 4520B are Dual synchronous 'UP' counters with 4-bit coded outputs. The 4518B is a Dual decade counter with BCD outputs, and the 4520B is a Dual hexadecimal (divide-by-16) counter with 4-bit binary outputs. The ICs have similar functional diagrams, as shown in *Figures 11.43* and *11.44.*

An unusual feature of these counters is that they can be clocked on either the positive (rising) or the negative (falling) edge of the clock signal. For positive-edge clocking, feed the clock to the CLK terminal and tie the ENABLE terminal high. For negative-edge clocking,

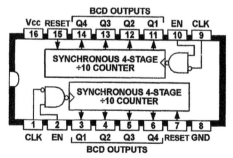

Figure 11.43 Functional diagram of the 4518B Dual synchronous decade counter with BCD outputs

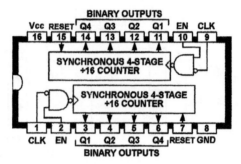

Figure 11.44 Functional diagram of the 4520B Dual synchronous 4-bit binary counter

feed the clock to the ENABLE terminal and tie the CLK terminal low. The counters are cleared by a high level on their RESET pins, and clear asynchronously.

Note that these counters do not have a 'carry' output; to cascade (ripple clock) counter stages, the negative-edge clocking feature must be utilized, as shown in *Figure 11.45*. Here, the Q4 output of each counter is fed to the ENABLE input of the following stage, which must have its CLK terminal tied low.

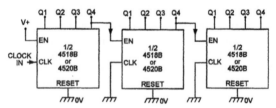

Figure 11.45 Method of cascading 4518B or 4520B counters for ripple operation

12   Special Counter/dividers

The last two chapters looked at D-type and JK flip-flops and showed how they are used in various 'up'-counting counter/divider ICs. The present chapter rounds off this subject by looking at some of the special types of digital counter/divider ICs that are available, including 'presettable' and 'down'- and 'up/down'- counting types.

Directional Control

Chapter 10 explained how D-type and JK flip-flops work and described how a D-type can be made to act as a binary divider (divide-by-2) stage by connecting its D and NOT-Q terminals together as shown in *Figure 12.1(a)*, and a JK-type can be made to act in the same way by tying its J and K terminals to logic-1 as shown in *Figure 12.1(b)*. Note from the symbols of these diagrams that the D-type's actions are triggered by the rising edges of the clock signals, and the JK's are triggered by falling edges.

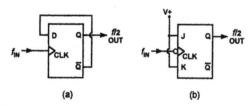

(a)   (b)

Figure 12.1   General circuit diagram of *(a)* D-type and *(b)* JK divide-by-2 stages; the D-type uses rising-edge triggering; the JK is falling-edge triggered

Figure 12.2 shows how D-type or JK divider stages can be cascaded to make ripple-mode binary counters in which the output of the first stage is used to clock the input of the second stage, and so on for however many stages there are. In rising-edge triggered D-type circuits the clock pulses are taken from NOT-Q outputs, as shown in *(a)*, but in falling-edge triggered JK circuits they are taken from Q outputs, as in *(b)*. Both circuits

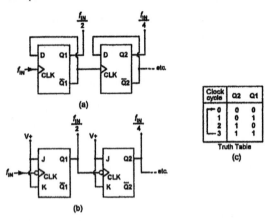

Figure 12.2   Basic ways of using *(a)* D-type and *(b)* JK stages in up-counting ripple modes

have the same Q-output Truth Table, which for a 2-stage ripple counter is as shown in *(c)*; if you compare this Table with that of *Figure 11.1* (in Chapter 11) you'll notice that the 4-step sequential binary coded outputs of *Figure 12.2* correspond with normal Decimal (BCD) coding, and run 0–1–2–3–0–, etc as the clock goes through cycles 0–1–2–3–0–, etc.

Thus, the *Figure 12.2* type of counter gives a sequentially upwards-counting clocking cycle, which repeatedly runs from 0 to 3 in a 2-bit counter, or 0 to 7 in a 3-bit counter, and so on. Consequently, all circuits of this basic type are known as 'up' counters, irrespective of their bit count or whether they give a synchronous or asynchronous type of clocking action, etc.

The counting action of a ripple counter can be reversed, so that it counts downwards rather than upwards, by using the basic connections shown in *Figure 12.3*, in which the clocking output of each stage is taken from the complement of that used in *Figure 12.2* (i.e. from Q rather than NOT-Q, or vice versa). A 2-bit counter of this type generates the Truth Table shown in *Figure 12.3*(c), in which the sequential BCD coding runs 3–2–1–0–3–, etc. as the clock goes through cycles 0–1–2–3–0–, etc. This type of counter gives a sequentially downwards-counting clocking cycle, which repeatedly runs from 3

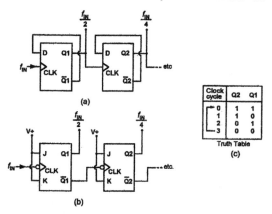

Figure 12.3 Basic ways of using *(a)* D-type and *(b)* JK stages in down-counting ripple modes

to 0 in a 2-bit counter, or 7 to 0 in a 3-bit counter, and so on. Consequently, all circuits of this basic type are known as 'down' counters, irrespective of their bit count or whether they give a synchronous or asynchronous type of clocking action, etc.

A ripple counter can be configured to count in either direction by fitting it with gate-controlled clock-source options, as shown in the JK example of *Figure 12.4*. Here, when the COUNT-UP line is biased to logic-1 and the COUNT-DOWN line is biased to logic-0 the G1 and G4 AND gates are enabled and pass 'Q' clock signals to FF2 and FF3 (etc.), but gates G2 and G5 are disabled and block the NOT-Q signals; the circuit thus acts like that of *Figure 9.2(b)* under this condition, and gives an up-counting action. But when the COUNT-DOWN line

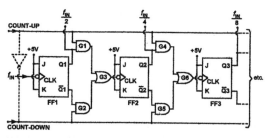

Figure 12.4 Basic JK-type up/down counter

is biased to logic-1 and COUNT-UP is at logic-0 the G1 and G4 gates are disabled and G2 and G5 are enabled and pass NOT-Q clock signals, and under this condition the circuit acts like that of *Figure 12.3(b)* and gives a down-counting action. All circuits of this basic type are known as 'up/down' counters, irrespective of their bit count or their precise form of construction, etc.

The *Figure 12.4* up/down circuit uses individual COUNT-UP and COUNT-DOWN control lines, which must always by connected in opposite logic states. The circuit can be modified for control via a single UP/DOWN input terminal by wiring an inverter between the two lines, as shown dotted in the diagram, so that the circuit gives 'up' counting when the upper line is biased to logic-1, and 'down' counting when it is biased to logic-0. Some up/down counters use two clock lines, one for up counting and the other for down counting; these counters are internally similar to *Figure 12.4* but have the clock and direction-control lines effectively combined via logic networks that control the clock feed to FF1 as well as all the other flip-flop stages.

Thus, the electronics engineer has many options when designing modern counter/divider circuits. The usual option is to use a conventional synchronous or asynchronous up-counting IC, and most of the finer points of this subject are covered in Chapter 11. There are also the options of using 'down' or 'up/down' counters; both of these options are dealt with later in the present chapter, but first it is necessary to look at yet another option, that of using 'programmable' counter/divider ICs.

'Programmable' Counter Basics

Many 'counter' ICs can be used in both counting and dividing applications, and are thus known as counter/divider ICs. *Figure 12.5*, for example, shows how two ordinary BCD divide-by-10 stages can be used as a divide-by-100 circuit that *(a)* gives a stable output frequency of 1kHz if fed with a stable 100kHz clock signal, or *(b)* produces 27 output pulses (cycles) when

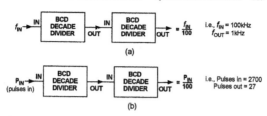

(a)

(b)

Figure 12.5 Two decade dividers used to give *(a)* frequency division or *(b)* pulse-count division

clocked via 2700 randomly timed input pulses. Note that neither of these circuits gives any visual output information (via 7-segment displays, etc.).

Figure 12.6 shows, in block diagram form, how the *Figure 12.5(b)* pulse cycle divider circuit can be converted into a pulse *counter* by simply fitting it with digital readout circuitry, so that the user can actually see the results of the divide-by-100 action. Thus the basic difference between a counter and a divider is one of usage; a counter usually has a visual readout, and a divider has not.

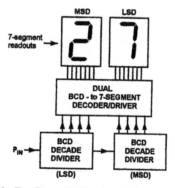

Figure 12.6 The *Figure 12.5(b)* pulse-count divider can be used as a pulse counter by fitting it with a digital readout facility

Most conventional 'up' counters have a RESET facility that enables all Q outputs to be set to zero at any time, thus giving a BCD '0' output. This facility is – as already described – also useful in enabling the counter's divide-by figure to be preset to any desired value, N, by connecting the outputs back to the RESET terminal so that the outputs reset to the BCD '0' state on the arrival of every *N*th clock pulse, but a weakness here is that the IC has to be hard-wired to give a specific divide-by figure, and this sometimes involves the use of external gating circuitry, etc. One way around this snag is to provide the IC with an additional PROGRAMMING control that enables its outputs to be set to any desired binary values when the control is activated.

Figure 12.7 illustrates the basic idea behind the programmable counter. Here, any desired 4-bit binary code can be applied to the IC's four 'P' terminals, and the IC's 4-bit output is forced to agree with this code whenever the PRESET control is activated. In practice, this type of IC may be known as a 'programmable' or 'presettable' counter, its input facilities may be named PRESET or

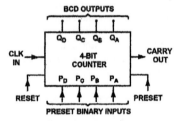

Figure 12.7   The Q outputs of a programmable counter can be forced into preset binary states via a PRESET control

PARALLEL LOAD or JAM controls, and the controls may be activated by a logic-0 or logic-1 input or by rising or falling clock edges, etc., depending on the individual device and its manufacturer. In all cases, however, these devices operate in the basic way just described.

Figure 12.8 shows one basic way of using the PRESET control to make a divider with any desired whole-

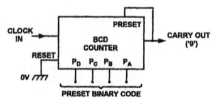

Figure 12.8   Basic way of using the PRESET facility of a programmable *divider*

number divide-by value from 2 to 9. Assume here that the PRESET terminal is active-high and is shorted to the CARRY output as shown; this output goes high on the arrival of each decimal-9 clock pulse and thence sets the IC's Q outputs to the preset state. Thus, if the DCBA preset inputs are set to '0000' the IC will go through a 0-1–2–3–4–5–6–7–8–0–, etc. counting cycle and thus give a divide-by-9 action, but if they are preset to '0010' (decimal-2) the IC will go through a 2–3–4–5–6–7–8–2–, etc. counting cycle and thus give a divide-by-7 action, and so on. This type of divider thus goes through X-to-8 counting cycles, where X is the number set on the DCBA preset inputs; note that this action is useful in a divider, but of little value in a decade counter (in which counting usually starts from zero).

Figure 12.9 shows one basic way of using the PRESET function in a counter. Here, the 7-segment readout displays the current preset BCD number while the PRESET control is activated, and the counter then counts up from that number when the PRESET control is deactivated; the BCD inputs may be derived from special switches, or may be taken from the BCD outputs of a slowly clocked up/down counter, etc. This latter technique is of special value in time-setting electronic clocks, etc.

Yet another use of the PRESET facility is as a master RESET control if the normal RESET is in permanent

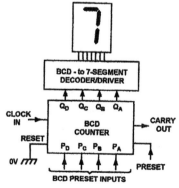

Figure 12.9 Basic way of using the PRESET facility of a programmable *counter*

use as a divide-by controller; in this case the preset binary code is simply set to '0000'.

Several popular 'up'-counter ICs are provided with programmable PRESET facilities. Among TTL types are the 74LS160, 74LS162 and 74LS196 decade counters, and the 74LS161, 74LS163 and 74LS197 4-bit binary counters; the most popular '4000'-series CMOS IC of this type is the 4018B (see Chapter 10). Most modern 'down'- and 'up/down'-counter/divider ICs are provided with programming facilities, and several examples of these are given in later sections of this chapter.

'Down' Counters

All 'down' counters should *ideally* have the basic facilities shown in *Figure 12.10*. Namely, they must be programmable (presettable) and have a special output that activates when the 'zero' count is reached, plus an input that inhibits the clocking action when activated. In the diagram, PRESET and INH (clock inhibit) are assumed to be active-high, and the ZERO OUT terminal goes high only when ZERO count is reached. In the following text and diagrams only decade (rather than binary) versions of these devices are considered, and the abbreviation 'PDDC' is used to indicate a programmable decade down counter.

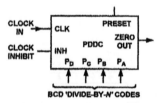

Figure 12.10 Basic features of a programmable decade down counter (PDDC)

A PDDC can be used as a simple decade divider by connecting it as shown in *Figure 12.11*, with its PRESET and INH controls, etc., grounded, so that the counter repeatedly cycles through its basic BCD count,

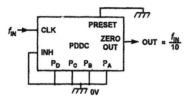

Figure 12.11   PDDC connected as a simple decade frequency divider

from 9 to 0 and then back to 9 again, and so on. The output, taken from the ZERO OUT terminal, goes high for one full clock cycle in every ten.

Figure 12.12 shows how to use a PDDC in its most important mode, as a programmable frequency divider. Here, the divide-by-N code is applied to the preset terminals and PRESET is controlled by the ZERO OUT terminal. Suppose that at the start of the count the BCD number 4 has been preset into the counter. On the arrival of the first clock pulse the counter decrements to 3, on the second pulse to 2, on the third to 1, and on the fourth to 0, at which point the ZERO OUT terminal goes high and presets the BCD number 4 back into the counter, so the whole sequence starts over again and ZERO OUT goes back low. Thus, the PDDC repeatedly counts by the number (4) set on the preset inputs, and the output (from the ZERO OUT terminal) takes the form of a narrow pulse with a width of a few tens of nanoseconds.

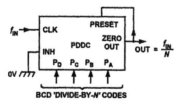

Figure 12.12   PDDC connected as a programmable frequency divider

The really important thing about the *Figure 12.12* circuit is that it automatically divides by whatever BCD number is set on the preset terminals (compare this action to that of the programmable 'up' counter of *Figure 12.8*, in which the preset BCD number is complexly related to the divide-by number). This feature is of special importance in what are known as 'decade cascadable' counter/divider applications, in which the overall division values are easily and directly programmable; *Figures 12.13* and *12.14* illustrate the salient points of the subject.

Figure 12.13   When conventional counters are cascaded they give a final output equal to the product of the individual division values

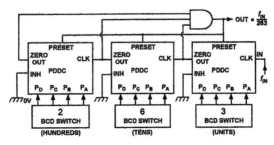

Figure 12.14   When PDDCs are wired in the decade cascaded mode they give a final 'divide-by' value equal to the *sum* of their individual *decade* values

Figure 12.13 shows a conventional circuit in which three normal divider stages, with divide-by values of 2, 6 and 3, are cascaded to give an overall division value of 36, i.e. equal to the *product* of the individual divide--by values. *Figure 12.14*, on the other hand, shows what happens when PDDCs with divide-by values of (reading from left to right) 2, 6 and 3 are wired in the decade cascadable mode; in this case the overall divide-by value is equal to the *sum* of the individual *decade* values, and (since the decades are graded in 'hundreds', 'tens', and 'units') equals 263, or whatever other 3-digit number is set up on the PRESET switches. The circuit operates as follows.

Note in *Figure 12.14* that the ZERO OUT signals of the three PDDCs are ANDed and used to drive the PRESET and OUT lines, which thus becomes active only when all three ZERO OUT signals coincide. With this point in mind, assume that at the start of the count cycle the

BCD number 263 is loaded into the counters as shown. For the first few counts in the cycle the unit's PDDC counts from 3 down to 0 and then goes into the normal 9-to-0 decade down-counting mode, passing a clock pulse on to the 'tens' PDDC each time the 0 state is reached. Thus, the 'tens' PDDC receives its first clock pulse after three input cycles and counts down from 6 to 5, but from then on is clocked down one step for every ten input cycles, until its own count falls to zero, at which point it passes a clock pulse on to the hundreds counter and simultaneously goes into the decade down-counting mode. One hundred input cycles later the 'hundreds' PDDC receives another clock pulse and its own count falls to zero; one hundred input cycles after that (on the 263rd count of the cycle) it receives a third clock pulse, and at that instant the ZERO OUT signals of all three PDDCs are active, so the AND gate activates the PRESET line and loads the BCD number 263 back into the counters, and the whole sequence starts over again.

Thus, the *Figure 12.14* circuit repeatedly divides by 263 or whatever other 3-decade number is programmed in, and produces a narrow output pulse (from the AND gate) on completion of each 'divide-by-263' counting cycle. This output pulse is only a few tens of nanoseconds wide (the width is dictated by the circuit's propagation delays), but is wide enough to trigger digital elements such as counters or monostables, etc.

PDDC Applications

Decade cascaded PDDC circuits of the *Figure 12.14* type have important practical applications in frequency synthesis, programmable counting, and programmable timing. In frequency synthesis the PDDCs are wired as a programmable frequency divider and used in conjunction with a phase-locked loop (PLL) as shown in *Figure 12.15*. Here, the output of a wide-range voltage-controlled oscillator (VCO) is fed, via the programmable divide-by-N counter, to one input of a phase detector, which has its other input taken from a crystal-

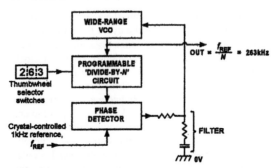

Figure 12.15 A programmable divide-by-*N* circuit can be used in conjunction with a PLL to make a precision programmable frequency synthesizer

controlled reference-frequency generator. The phase detector produces an output voltage proportional to the difference between the two input frequencies; this voltage is filtered and fed back to the VCO control in such a way that the VCO automatically self-adjusts to bring the variable input frequency of the phase detector to the same value as the reference frequency, at which point the PLL is said to be 'locked'.

Note that, when the PLL is locked, the VCO's output frequency is *N* times that on the variable input of the phase detector, and is thus *N* times that of the reference generator, e.g. if $N = 263$ and $f_{REF} = 1kHz$, $f_{OUT} = 263kHz$, and has crystal precision. Thus, this circuit can be used to generate precise output frequencies that are variable in 1kHz steps via the 3-decade thumbwheel selector switches.

Figure 12.16 shows, in basic form, how a single PDDC can be used as a simple counting circuit. Here, the INH

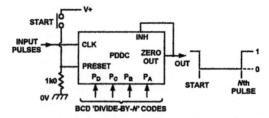

Figure 12.16 PDDC used as a programmable down counter

terminal is connected directly to ZERO OUT, so that the PDDC's clocking action is inhibited when the PDDC is in the '0' counting state; normally, the circuit is locked into this state, with its output at the logic-1 level. Suppose now that the BCD number 6 is preset via the START button; the output immediately switches low, and on the arrival of each clock pulse the PDDC counts down one step until finally, on the arrival of the sixth pulse, the ZERO OUT terminal goes high again and activates INH, causing any further pulses to be ignored. The count sequence is then complete, but can be restarted via the START button.

Figure 12.17 shows how the basic circuit can be turned into a useful 2-decade unit that can be programmed to count by any number up to 99 and has an output that goes high for the full duration of the counting cycle. This type of circuit is useful in (for example) controlling automatic packing machines in applications where *N* objects have to be loaded into each container, but the *N* value is often varied. In such an application, the object feeder must generate a clock pulse each time it feeds an object into the container, and must be so arranged that it directs its feed to the next container when the first one is registered 'full'.

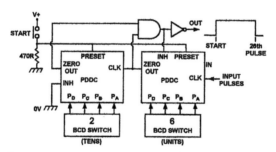

Figure 12.17 2-decade programmable down counter, set for count-26 operation

Finally, *Figure 12.18* shows, in basic form, how the simple *Figure 12.16* circuit can be made to act as a programmable timer with an output that goes high as the START button is pressed but then goes low again a preset time later. Here, the clock signal (which ideally should be synchronized with the START signal) is taken from a

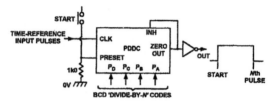

Figure 12.18   PDDC connected as a programmable timer

time-reference source (e.g. 1 pulse per second or minute, etc.). This basic circuit can be expanded to two decades by using connections similar to those of *Figure 12.17*.

Practical PDDC circuits can be built by using either dedicated programmable down-counter ICs, or by using up/down counters wired into the down-counting mode. No dedicated programmable 'down' counters are available in TTL form, but several are available in CMOS form, the best known examples being the 4522B and 4526B 'Single' and the 40102B (or 74HC40102) and 40103B (or 74HC40103) 'Dual' ICs, which are described in the next two sections of this chapter.

The 4522B and 4526B ICs

The best known family of CMOS programmable cascadable down counters comprise the 4522B (BCD decimal) and the 4526B (binary) 4-bit ICs, which have the functional diagrams shown in *Figures 12.19* and *12.20*. Each of these ICs contains a 4-bit down counter that has its Q1 to Q4 outputs externally available and can be synchronously reset to zero by taking the MASTER RESET pin high, and which can be loaded with a 4-bit 'divide-by' code (applied to the P1 to P4 pins) via the pin-3 LOAD control. The counter clock (CLK) signal can be inhibited via the INH terminal. An AND gate is built into the ZERO output line, so that the ZERO output can only go high if the CASCADE FEEDBACK (CF) terminal is also high, thus enabling cascading to be achieved without the use of external gates.

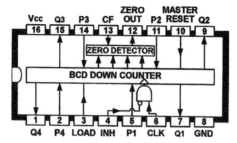

Figure 12.19 Functional diagram of the 4522B programmable 4-bit BCD (decade) down counter

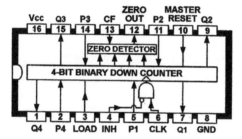

Figure 12.20 Functional diagram of the 4526B programmable 4-bit (binary) down counter

The 4522B decade down counter can be used as a PDDC in the same basic ways as shown in *Figures 12.11* to *12.18*, except that the MASTER RESET pin is normally grounded and the AND gate is already built into the IC. When the IC is used alone, the CASCADE FEEDBACK (CF) terminal must be tied high to enable the ZERO output. When cascading two or more ICs, tie the ZERO output of the MSD package to the CF terminal of the (MSD – 1) package, repeating the process on all less-significant dividers except the first. *Figure 12.21* shows practical connections for making a 2-stage programmable down counter, and *Figure 12.22* shows the connections for a 2-stage programmable frequency divider, using either 4522B or 4526B ICs.

When using these ICs, note that all unused inputs (including PRESETs) must be tied high or low, as appropriate, and that the outputs of all internal counter stages are available via the Q terminals, enabling the counter states to be decoded via external circuitry.

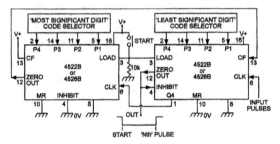

Figure 12.21 2-stage programmable down counter, using 4522B or 4526B ICs

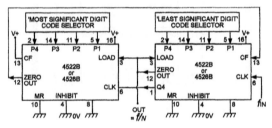

Figure 12.22 2-stage programmable frequency divider, using 4522B or 4526B ICs

The 40102B and 40103B ICs

The 40102B and 40103B are another popular family of CMOS programmable cascadable down-counter ICs, but in this case each device houses a cascaded pair of presettable 4-bit down counters, with only the NOT-ZERO output (which goes low under the zero count condition) of the counters externally available; Q outputs are not provided. *Figure 12.23* shows the functional diagram of the 40102B (and its '74HC'-series counterpart, the 74HC40102), which functions as a 2-decade BCD-type down counter, and *Figure 12.24* shows the functional diagram of the 40103B (and 74HC40103), which functions as an 8-bit binary down

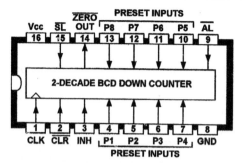

Figure 12.23 Functional diagram of the 40102B (or 74HC40102) 2-decade BCD down counter

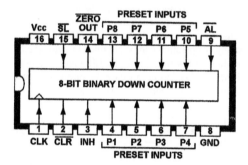

Figure 12.24 Functional diagram of the 40103B (or 74HC40103) 8-bit binary down counter

counter. Both types clock down on the positive transition of the CLK signal. *Figure 12.25* shows the Truth Table that is common to both ICs.

Codes that are applied to the eight PRESET pins of the ICs can be loaded asynchronously by pulling the NOT-

TRUTH TABLE

CONTROL INPUTS				'LOAD' MODE	ACTION
CLR	AL	SL	INH		
1	1	1	1	SYNC	INHIBITS COUNTER
1	1	1	0	SYNC	COUNT DOWN
1	1	0	X	SYNC	LOAD ON NEXT CLOCK PULSE
1	0	X	X	ASYNC	LOAD ASYNCHRONOUSLY
0	X	X	X	ASYNC	CLEAR TO MAXIMUM COUNT

NOTE: X = Don't care

Figure 12.25 Truth Table of the 40102B and 40103B (and 74HC40102 and 74HC40103) down-counting ICs

AL pin low, or synchronously (on the arrival of the next
CLK pulse) by pulling the NOT-SL pin low. When the
NOT-CLR input is pulled low the counter
asynchronously clears to its *maximum* count. When the
inhibit (INH) control is pulled high it inhibits both the
clock counting action and the NOT-ZERO output action,
thereby acting as a 'carry-in' terminal in cascaded
applications.

Figures 12.26 to *12.29* show four basic ways of using this
family (or their '74HC'-series counterparts) of presettable
down counters. *Figure 12.26* shows the connections for
making a programmable 8-bit or 2-decade timer or down
counter, and *Figure 12.27* shows the circuit of a
programmable frequency divider. The latter circuit gives
divide-by-$(N + 1)$ operation, its output going low for one
full clock cycle under the 'zero count' condition; true
divide-by-N operation can be obtained by tying the NOT-
SL pin high and wiring the NOT-ZERO output (pin-14) to
NOT-AL (pin-9), but in this case the output pulses have
widths of only a few hundred nanoseconds.

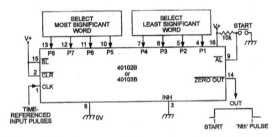

Figure 12.26   Programmable timer using a 40102B or 40103B
(etc.) IC

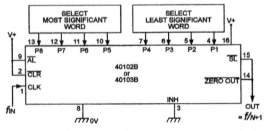

Figure 12.27   Programmable frequency divider (divide-by-$(N + 1)$)

Finally, *Figures 12.28* and *12.29* show the basic connections that are used to cascade 40102B or 40103B stages in large-bit programmable applications. The *Figure 12.28* connections give ripple operation, and the *Figure 12.29* connections give fully synchronous operation (for high-speed applications).

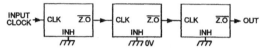

Figure 12.28 Basic way of ripple cascading 40102B or 40103B counters

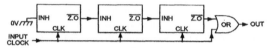

Figure 12.29 Basic way of synchronously cascading 40102B or 40103B counters

'Up/down' Counters

'Up/down' counters (UDCs) are the most versatile of all counter types. They are invariably programmable and synchronous in operation, are available in both BCD decade and 4-bit binary counting forms, and in most cases have the basic facilities shown in *Figure 12.30*, i.e. they have PRESET inputs and a full set of Q outputs, can be set to clock up or down via a single UP/NOT-DOWN terminal, have a clock inhibit (INH) facility, and have a TC OUT output that becomes active

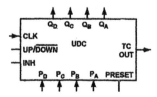

Figure 12.30 Basic features of a conventional up/down counter (UDC)

when the counter reaches its Terminal Count ('0' in down-counting mode, '9' in decade up-counting mode, etc.). The 40192B and 74LS/HC192 (BCD decade) and 40193B and 74LS/HC193 (4-bit binary) UDCs differ from this norm in that they use separate 'up' and 'down' clocks and have no master INH facility, but are otherwise similar. Individual UDC IC types differ mainly in their pin terminology (the INH terminal may, for example, be named NOT-ENABLE or CARRY IN, etc.) and in their details of use, i.e. INH or TC OUT may be active-high on one IC type and active-low on another.

Because of their versatility and consequent high sales volumes, up/down counters are often available at lower cost than conventional types, and can thus be used instead of normal or programmable 'up' or 'down' counters in a wide range of applications, as well as being invaluable in add/subtract and up/down differential counting applications, etc. *Figures 12.31* to *12.41* show a selection of different ways of using the basic *Figure 12.30* UDC; in these diagrams it is assumed that the INH, PRESET and TC OUT controls are active-high, and that the IC counts up when the UP/NOT-DOWN control is biased high, and down when it is biased low.

Figure 12.31 shows a decade up/down counter used as a simple decade up counter/divider, in which PRESET can be used as a RESET control that forces the outputs into the 0000 state when it is taken high. *Figure 12.32* shows the UDC used as a programmable up divider of the *Figure 12.8* type, and *Figure 12.33* shows it wired as a programmable frequency divider of the far more useful

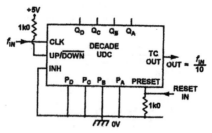

Figure 12.31 A decade UDC used as a simple decade up counter/divider

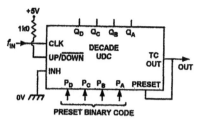

Figure 12.32 A decade UDC used as a programmable up divider

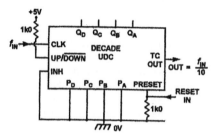

Figure 12.33 A decade UDC used as a 'down' (PDDC) programmable frequency divider

Figure 12.12 down-counting PDDC type, in which the divide-by value equals the BCD value set on the programming terminals; this basic down-counter circuit can easily be used in the decade cascaded modes shown in *Figures 12.14* and *12.17*, etc.

Figure 12.34 shows, in basic form, how a 4-bit Binary UDC can be used as (*a*) a digital-to-analogue converter (DAC) driver or (*b*) a multi-line selector. The DAC

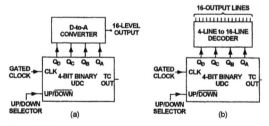

Figure 12.34 Basic ways of using a binary UDC as (*a*) a DAC driver or (*b*) a multi-line selector

driver produces 16 selectable output voltage levels when driven by the 4-bit Binary UDC as shown, or 256 levels if two UDCs are cascaded to give 8-bit DAC drive; these output voltage levels can easily be used to control sound levels or lamp brightness, etc., via suitable adaptor circuitry. The *(b)* circuit can be used as a multi-line selector, in which only one of the 16 output lines is active (usually active-low) at any one time, by using the UDC to drive a 4-line to 16-line decoder as shown, or can be used like a single-pole 16-way switch by using it to drive a CMOS analogue switch IC such as the 4067B, etc.

Figure 12.35 shows, in basic form, how a BCD decade UDC can be used as *(a)* a 4-bit BCD code generator or *(b)* a multi-line selector. In the former case the BCD code can be used to drive PRESET inputs and/or 7-segment digital displays, etc., and the circuit is thus useful in the time setting of clocks and presetting of counters, etc. The *(b)* circuit can be used in the same ways as the multi-line selector of *Figure 12.34*, but gives only ten outputs; its BCD output can, however, be used to simultaneously drive a digital display that shows the prevailing output number. Also, by using multiplexing or ANDing techniques, two of these basic circuits can be cascaded to make up to one hundred individual outputs available.

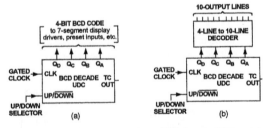

Figure 12.35 Basic ways of using a BCD decade UDC as *(a)* a 4-bit BCD code generator or *(b)* a multi-line selector

Figure 12.36 shows typical connections that may be used in practical versions of the *Figure 12.34* or *12.35* gate-clocked circuits. Here, a 555 IC is used in the astable mode as a clock-waveform generator, and is normally disabled but can be gated on by grounding its

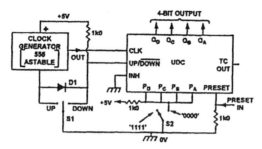

Figure 12.36 Typical connections of a gate-clocked free-ranging UDC circuit

negative supply line via centre-biased toggle switch S1, which also controls the UDC's up/down direction. The UDC's preset inputs are shown configured so that they give codes of 0000 (= BCD '0') when S2 is closed or 1111 (= terminal count in Binary up-counting mode) when S2 is open, enabling a Binary UDC to be instantly set to either end of its counting range.

Note that the *Figure 12.36* circuit gives a 'free-ranging' clocking action, i.e. when it reaches the terminal count in a clocking cycle it automatically jumps back to the 'start' count on the arrival of the next clock pulse. A popular alternative to this is an 'end-stopped' counting action, in which counting automatically ceases when the terminal count is reached, and can only be restored by reversing the count direction, i.e. so that a lower number can only be reached by clocking down, and a higher number can only be reached by clocking up. If the UDC's INH and TC OUT terminals have the same active states (both active-high or active-low) this action can be obtained by simply shorting these two terminals together as shown in *Figure 12.37,* but if they have opposite active states an inverter stage must be wired between TC OUT and INH as indicated in the diagram (the UDC's 'preset' connections are not shown in the diagram).

The basic *Figure 12.36* free-ranging circuit can be expanded to give an 8-bit output by cascading it with another UDC, with both INH terminals grounded and with both UP/NOT-DOWN terminals shorted together, etc., and with the TC OUT of the first UDC providing

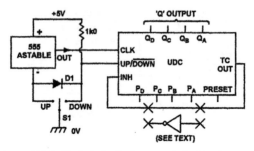

Figure 12.37   Typical circuit of a gate-clocked 'end-stopped' UDC circuit (PRESET connections not shown)

the CLK signal of the second IC. The procedure for expanding the end-stopped version is a little more complex, as shown in *Figure 12.38*. In this case both UP/NOT-DOWN terminals are again shorted together, and the TC OUT of UDC1 provides the CLK signal for UDC2, but the TC OUTs of both counters are ANDed and fed to INH of UDC1, thus providing bi-directional end-stopping, and INH of UDC2 is grounded. Note in these expanded circuits that the first UDC (UDC1) provides the four least-significant bits of the 8-bit output.

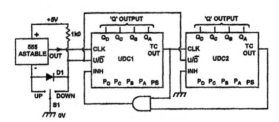

Figure 12.38   Basic method of expanding the *Figure 12.37* end-stopped UDC circuit to give an 8-bit output

'Add/subtract' Circuits

A UDC can be made to perform add/subtract actions by simply switching it to the up mode for addition or the down mode for subtraction and clocking in one input pulse per add or subtract unit. This type of action is useful in applications such as car park monitoring, etc., and *Figure 9.39* shows the basic circuit of a monitor that keeps track of the number of cars in a car park with a maximum capacity of 99 vehicles. Assume here that the car park's entry/exit is fitted with a system that generates a single clock pulse and a logic-1 level whenever a vehicle enters the site, and a single pulse and a logic-0 level whenever a vehicle leaves the site, and is fitted with fool-proof logic circuitry that ensures that only one of these sets of states can occur at any given moment. Also assume that the circuit uses two UDCs connected in the end-stopped counting fashion shown in *Figure 12.38*, and thus cannot count below zero or above 99; the basic circuit operates as follows.

At the start of each day the car park attendant operates PRESET switch S1, setting the digital readouts to zero. The readout then increments by one count each time a vehicle enters the car park or decrements by one count each time a vehicle leaves, thus giving a running count of the number of vehicles parked; when this number reaches 99, the TC OUT and UP/NOT-DOWN terminals are both at logic-1 and are ANDed and used to activate the car park's 'FULL' sign (note that this AND gate stops the 'FULL' sign activating when TC OUT goes high at zero count in the down mode).

The basic *Figure 12.39* circuit has two obvious defects. First, it has no facility for correcting counting errors, such as occur if the attendant activates the PRESET switch without realizing that a number of vehicles

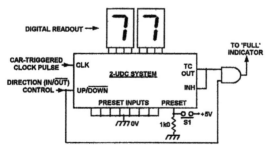

Figure 12.39 Simple car park 'number-of-cars-parked' indicator (99 maximum)

remained locked in overnight. Second, it only activates the 'FULL' sign correctly if the car park has a maximum capacity of 99 vehicles. *Figure 12.40* shows how both of these defects can be overcome. Here, S2 is a biased 2-pole toggle switch that is normally set to AUTO; any start-of-the-day counting errors can be rectified – after setting the count to zero via S1 – by moving S2 to the MANUAL position and feeding in manually triggered clock UP pulses via S3 until the reading is correct. The second defect is overcome by a set of gates connected as a MAXIMUM-COUNT DETECTOR, which gives a logic-1 output and activates the 'FULL' sign only when the system's two 4-bit BCD output codes correspond with the car park's maximum capacity.

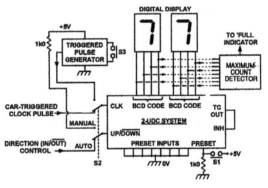

Figure 12.40 Improved car park 'number-of-cars-parked' indicator can show any 'FULL' value up to 99 maximum

An alternative – and probably better – approach to the 'car park' problem is to have a system that displays the available number of parking spaces, rather than the number of parked cars. *Figure 12.41* shows a basic circuit of this type. Here, the car park's entry/exit system is made to generate a low (down-count) level when a car enters, and a high (up-count) level when a car leaves. When the S1 PRESET switch is operated it loads the car park's BCD 'maximum number of spaces' code into the system, and this number appears on the display; any real-life errors in this 'spaces' reading can be corrected by setting S2 to MANUAL and operating

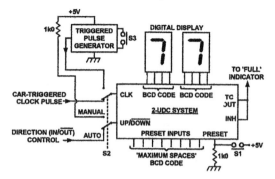

Figure 12.41 Car park 'spaces available' indicator can show up to 98 maximum spaces

S3. The 'spaces' readout then decrements by one count each time a car enters the car park or increments by one count each time a car leaves, thus giving a running count of the number of available spaces; when this number falls to zero the TC OUT terminal goes high and activates the car park's 'FULL' sign (note that this sign will also activate if the 'spaces' count *rises* to 99, so this circuit functions correctly up to a maximum 'spaces' value of only 98, unless suitably modified; the maximum capacity can easily be expanded to 998 spaces, by basing the design on 3-UDC system).

'Up/down' Counter ICs

Several different up/down-counter ICs are readily available in CMOS and TTL forms. In ordinary CMOS, the 4029B, 4510B, 4516B, 40192B and 40193B are very popular. The 4029B is unusual in that it has a control terminal that enables it to count in either decade or binary mode; the 4510B decade and 4516B binary counters are single-clock ICs with identical pin notations, and the 40192B decade and 40193B binary counters are dual-clock ICs with identical pin notations; four of these ICs are also available in the '74HC'-series as the 74HC4510, 74HC4516, 74HC40192, and 74HC40193.

In the '74LS'-TTL series, the most popular up/down counters are the 74LS190 and 74LS192 'decade' types and the 74LS191 and 74LS193 'binary' types. Of these, the '190' and '191' are single-clock ICs with identical pin-outs and control functions, and the '192' and '193' are dual-clock types with identical pin-outs and control functions. These four counters are also available in '74HC'-series CMOS versions, as the 74HC190, 74HC191, 74HC192, and 74HC193.

All of the above ICs are described in greater detail in the remaining sections of this chapter.

The 4029B UDC IC

Figure 12.42 shows the functional diagram of the 4029B presettable up/down-counter IC, and *Figure 12.43* shows the basic usage circuit of this very versatile

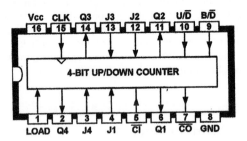

Figure 12.42   Functional diagram of the 4029B presettable up/down counter

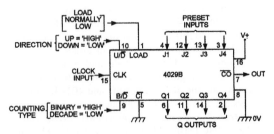

Figure 12.43   Basic usage circuit for the 4029B up/down counter

device, which acts as a binary counter when the B/NOT-D (pin-9) terminal is high, or as a decade counter (with BCD 'Q' outputs) when the terminal is low. The IC counts up when the U/NOT-D (pin-10) terminal is high, or down when the terminal is low (note, however, that this terminal should only be changed when the clock signal is high). The LOAD terminal is normally held low; it forces the outputs to immediately (asynchronously) agree with the binary code set on the 'J' inputs when it is taken high.

The IC's NOT-CI (pin-5) 'carry in' terminal is actually a clock disabling input, and is held low for normal clocking operation. When NOT-CI and LOAD are low, the counter shifts (up or down) one count on each rising edge of the clock signal. The NOT-CO (pin-7) 'carry out' terminal of the IC is normally high and goes low only when the counter reaches its maximum count in the up mode or its minimum count in the down mode (provided that the NOT-CI input is low).

The actions of the 4029B's NOT-CI and NOT-CO terminals are designed to facilitate fully synchronous clocking in multi-stage counting applications, as shown in the basic circuit of *Figure 12.44*, where all ICs are clocked in parallel and the NOT-CO terminal of each counter is used to enable the following one (at a 'one-tenth of' rate) via its NOT-CI terminal.

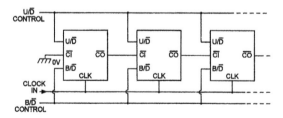

Figure 12.44   Basic way of wiring 4029Bs for synchronous parallel clocking

Numbers of 4029Bs can also be cascaded and clocked in the asynchronous ripple mode by using the basic connections shown in *Figure 12.45*, but in this case the Q outputs of different counters may not be glitch-free when decoded. Note in this diagram that the CLK and

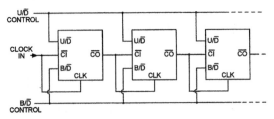

Figure 12.45   Basic way of wiring 4029Bs for asynchronous ripple clocking

NOT-CI terminals are joined together on each IC, to ensure that false counting will not occur if the U/NOT-D input is changed during a terminal count.

The 4510B and 4516B UDC ICs

The 4510B and 4516B are single-clock presettable up/down counters. The 4510B (and its 'HC' equivalent, the 74HC4510) is a 4-bit decade counter with BCD outputs, and the 4516B (and 74HC4516) is a 4-bit binary counter. The ICs have the functional diagrams shown in *Figures 12.46* and *12.47*, and have a RESET terminal which is held low for normal operation but sets all Q outputs to '0' (cleared) when biased high. The outputs can be forced to agree with the binary code preset on the 'P' terminals by applying a logic-1 level to the LOAD (pin-1) terminal. The ICs count up when the U/NOT-D terminal (pin-10) is high, and down when the

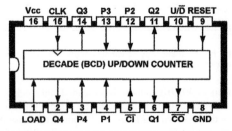

Figure 12.46   Functional diagram of the 4510B (or 74HC4510) decade up/down counter

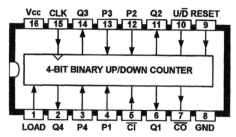

Figure 12.47 Functional diagram of the 4516B (or 74HC4516) binary up/down counter

terminal is low; the U/NOT-D input must only change when the clock signal is high. *Figure 12.48* shows the basic usage circuit that applies to this family of ICs.

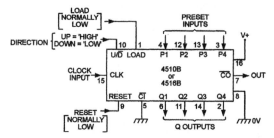

Figure 12.48 Basic usage circuit for the 4510B, 4516B, 74HC4510 and 74HC4516 up/down-counter ICs

The NOT-CI (pin-5) terminal is actually a clock disabling input, and is held low for normal operation. When NOT-CI (and RESET and LOAD) is low, the counter shifts one count on each rising edge of the clock signal. The NOT-CO (pin-7) terminal is normally high and goes low only when the counter reaches its maximum count in the up mode or its minimum count in the down mode (provided that the NOT-CI terminal is low).

Numbers of 4510Bs or 4516Bs (etc.) can be cascaded and clocked in parallel to give fully synchronous action by using the basic connections shown in *Figure 12.49*, or can be cascaded in the asynchronous ripple mode by using the basic connections of *Figure 12.50*, which ensures counting validity even if the U/NOT-D input is changed during a terminal count.

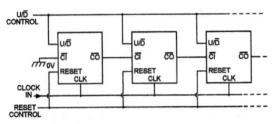

Figure 12.49   Basic way of wiring the 4510B/16B, etc., ICs for synchronous parallel clocking

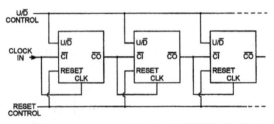

Figure 12.50   Basic way of wiring the 4510B/16B, etc., ICs for asynchronous ripple clocking

The 40192B and 40193B UDC ICs

The 40192B and 40193B are dual-clock presettable up/down counters. The 40192B (and its 'HC' equivalent, the 74HC40192) is a 4-bit decade counter with BCD outputs, and the 40193B (and 74HC40193) is a 4-bit binary counter. The ICs have the functional diagrams shown in *Figures 12.51* and *12.52*, and have a RESET terminal which is held low for normal operation but sets all Q outputs to '0' (cleared) when biased high. The outputs can be forced to agree with the binary code preset on the 'J' terminals by applying a logic-0 level to the NOT-LOAD (pin-11) terminal, which is normally biased high.

The counters have two clock lines, one controlling the up count and the other controlling the down count. Only one clock terminal must be used at a time; the unused input must be biased high. The counter shifts one count (up or down) on each rising edge of the used clock line. The NOT-CARRY (pin-12) and NOT-BORROW (pin-13) outputs of

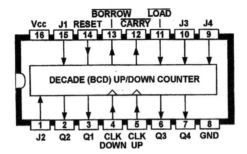

Figure 12.51 Functional diagram of the 40192B (or 74HC40192) dual-clock decade up/down counter

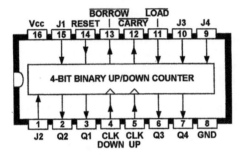

Figure 12.52 Functional diagram of the 40193B (or 74HC40193) dual-clock binary up/down counter

the counters are normally high, but the NOT-CARRY output goes low one-half clock cycle after the counter reaches its maximum count in the up mode, and the NOT-BORROW output goes low one-half clock cycle after the counter reaches its minimum count in the down mode.

Figure 12.53 shows the basic way of wiring the 40192B or 40193B (etc.) in cascaded multiple IC applications.

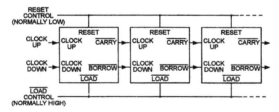

Figure 12.53 Basic way of cascading 40192B or 40193B (etc.) dual-clock up/down counters

The NOT-BORROW and NOT-CARRY outputs of each IC are simply connected directly to the CLK DOWN and CLK UP inputs respectively of the following ICs.

The 74LS/HC190 and 74LS/HC191 UDC ICs

Figure 12.54 shows the functional diagrams and pin-outs of the 74LS190 and 74LS191 single-clock up/down counters (and their fast CMOS equivalents, the 74HC190 and 74HC191). These two ICs have very similar active characteristics; namely, they trigger on the rising edge of the clock signal, and count up when pin-5 is low or down when pin-5 is high (assuming that INH is low and NOT-PRESET is high); the TC OUT terminal is normally low, but goes high for one clock cycle when the IC reaches its terminal count (0 in the down mode,

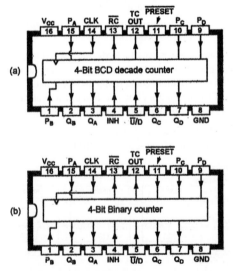

Figure 12.54 Functional diagrams of the *(a)* 74LS/HC190 decade and *(b)* 74LS/HC191 binary 4-bit up/down-counter ICs

or, in the up mode, 9 in the 74LS190 or 15 in the 74LS191). Note that both ICs have a terminal notated NOT-RC; this 'ripple clock' terminal is normally high but goes low on the falling edge of the terminal count and goes high again on the next clock rising edge; the terminal provides a clean clocking signal in multi-stage circuits, and is used as shown in *Figures 12.55* and *12.56*.

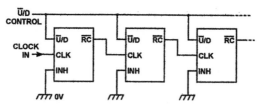

Figure 12.55 Basic ways of using the 74LS/HC190 or 74LS/HC191 as an *N*-stage ripple-clocked up/down counter (with NOT-PRESET at logic-1)

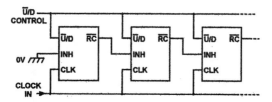

Figure 12.56 Basic ways of using the 74LS/HC190 or 74LS/HC191 as an *N*-stage synchronously clocked up/down-counter (with NOT-PRESET at logic-1)

Figure 12.55 shows the basic way of using the '190' or '191' as multi-stage up/down counters, using ripple clocking; in essence, the ripple clock signal of the first counter acts as the clock of the second stage, and the ripple clock of that acts as the clock of the third stage, and so on for however many stages there are. *Figure 12.56* shows how to use the counters in the fully synchronous clocking mode; in this case all ICs are clocked in parallel, but the ripple clock output of each stage controls the INH action of the next stage. Note that the NOT-PRESET control is not shown in these two circuits, but should be tied to logic-1 for normal clocked operation.

The 74LS/HC192 and 74LS/HC193 UDC ICs

Figure 12.57 shows the functional diagrams and pin-outs of the 74LS192 and 74LS193 dual-clock up/down counters (and their fast CMOS equivalents, the 74HC192 and 74HC193). These ICs have similar active characteristics; they trigger on the rising edge of the clock signal, and count up when the clock signal is applied to pin-5, or down when it is applied to pin-4; only one clock line must be used at a time, and the inactive one must be tied high. The CLK CARRY 'UP' and 'DOWN' outputs are normally high terminal count outputs that are useful in multi-stage ripple clocking applications; the 'UP' output goes low on the falling edge following the 'UP' terminal count, and returns high on the next rising edge; the 'DOWN' output similarly goes low, etc., on the 'DOWN' terminal count of '0'. The CLR

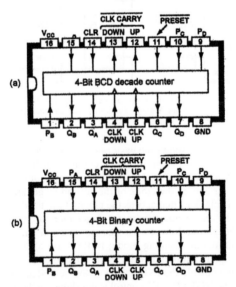

Figure 12.57 Functional diagrams of the *(a)* 74LS/HC192 decade and *(b)* 74LS/HC193 binary 4-bit dual-clock up/down-counter ICs

terminal is normally low, and sets all Q outputs to '0' when taken high; NOT-PRESET is normally high, and loads the PRESET data when taken low.

Figure 12.58 shows the basic way of using the '192' or '193' as multi-stage up/down counters, using ripple clocking; here, the 'CLK CARRY' 'UP' and 'DOWN' outputs of the first counter act as the 'UP' and 'DOWN' clocks of the second stage, and the 'CLK CARRY' outputs of that stage act as the clocks of the third stage, and so on for however many stages there are. Note that the CLR terminal is tied low for normal operation; the NOT-PRESET control isn't shown, but should be tied to logic-1 for normal clocked operation.

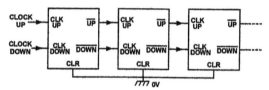

Figure 12.58 Basic ways of using the 74LS/HC192 or 74LS/HC193 as an *N*-stage ripple-clocked up/down counter (with NOT-PRESET at logic-1)

Counter Conversions

In most practical applications a dual-clock UDC has no intrinsic advantage over a single-clock type, or vice versa, and one type can easily be made to act like the other by using suitable input conversion logic circuitry. *Figure 12.59* shows a converter that makes a dual-clock

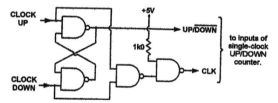

Figure 12.59 Converter gives dual-clock action on a single-clock up/down counter

counter act like a single-clock type (the circuit is shown set for 5V operation), and *Figure 12.60* shows a converter that enables a single-clock counter to be driven in the dual-clock mode; in multi-stage counters, these converters must be applied to the input(s) of the first stage only. If the counter's up/down 'active' levels are the reverse of those shown in the diagrams, simply reverse the input connections to the circuit in *Figure 12.59*, or reverse the output connections from the circuit in *Figure 12.60*.

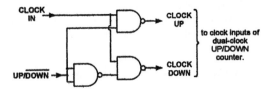

Figure 12.60   Converter gives single-clock action on a dual-clock up/down counter

13 Latches, Registers, Comparators and Converters

Data latches and shift registers are widely used members of the flip-flop family of devices, and are often used in conjunction with logic comparators and code converters. This chapter looks at LS TTL and CMOS versions of all of these devices and shows how to use them.

Data Latches

Flip-flop circuit elements of the SET–RESET, D-type, and JK types (see Chapter 10) are often called 'latches', because their outputs can be latched into either a logic-0 or logic-1 state by applying suitable input signals. JK and D-type latches are fairly versatile elements, and can be made to act as either data latches or as divide-by-2 circuits by suitably connecting their input and output terminals. A pure 'data latch', on the other hand, is an element that is built as a dedicated data latch and can be used for no other purpose; an element of this type acts as a simple memory that can store one 'bit' of binary data for an indefinite period; four such elements can store a complete 4-bit binary 'word'.

A data latch stores and outputs whatever logic level is applied to its 'D' (data) terminal when activated by a suitable 'store' command; the 'store' terminal may be either level or edge sensitive, and *Figure 13.1* shows the basic data latch symbols that apply in each case. Level-

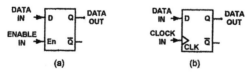

Figure 13.1 Symbols of *(a)* level-triggered and *(b)* edge-triggered data latches

triggered elements of the *Figure 13.1(a)* type are 'transparent' when the En (enable) terminal is high (i.e. the Q output follows the D input under this condition), but latch the prevailing D state into the Q output when En goes low; data can thus be latched by applying a brief positive pulse to the En terminal. Edge-triggered elements of the *Figure 13.1(b)* type are not transparent, and the Q output ignores the D input until a clock-pulse trigger edge is applied to the CLK input, at which point the latch stores and outputs the data and holds it until a new clock pulse arrives.

A 4-bit data latch can be built by connecting four 1-bit latches together as shown in *Figure 13.2*, with all En (or CLK) terminals wired in parallel so that all elements activate at the same time. Edge-triggered 4-bit data latches are not as useful, popular, or as readily available as level-triggered types, but can easily be built from JK or D-type flip-flop ICs such as the 74LS73, 74LS74, 4013B or 4027B, etc. Level-triggered 4-bit or greater transparent data latch ICs are readily available, at low cost.

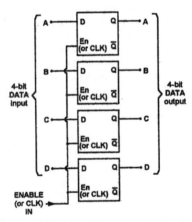

Figure 13.2 A 4-bit data latch is made from four 1-bit latches connected as shown

Level-triggered 4-bit data latches are widely used as temporary memories in digital display-driving applications, and may be used in either of the two basic ways shown in *Figure 13.3*. Assume here that the

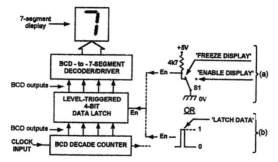

Figure 13.3 Basic ways of using a level-triggered 4-bit data latch in a display-driver application

decade counter shown is one of a cascaded chain of such counters, each of which has its BCD outputs fed to a 7-segment digital display via a 4-bit data latch and a decoder/driver IC, as shown. In simple counting applications the data latch may be used as in *(a)*, with its En terminal connected to the V+ rail via a 4k7 resistor so that it is biased high when S1 is set to the ENABLE position but is grounded when S1 is in the FREEZE position. Thus, in the ENABLE position the Latch is transparent, and the counter's state is instantly shown on the display (which may appear as a blur if fast counting is taking place), but when S1 is moved to the FREEZE position the data is immediately latched and the display is effectively frozen (i.e. changes in the counter's states are no longer displayed).

In the alternative control mode shown in *(b)*, the data latch's En terminal is fed with a timed chain of positive LATCH DATA pulses, and the display thus shows a grabbed snapshot of the counter's instantaneous state each time one of these pulses terminates, and displays it until the arrival of the next pulse's termination point. This technique is useful in cases where the display normally appears as a fast-changing blur; by feeding in the LATCH DATA pulses at (say) a ten-per-second (100ms) rate, the display can be strobed so that it appears in a varying but clearly readable form. This technique is widely used in digital frequency meters, which use the basic operating principle shown in *Figure 13.4.*

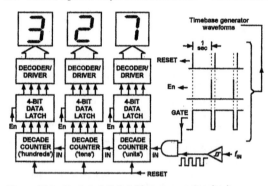

Figure 13.4 Basic 3-digit digital frequency meter circuit

The *Figure 13.4* circuit is that of a simple 3-digit frequency meter in which the test frequency is fed to the input of the counter chain via a 2-input AND gate, which is controlled via a timebase generator that produces a repeating timing cycle. At the start of this cycle the gate is closed, and a brief RESET pulse is fed to all three counters, which clear to zero; the gate then opens, and the counters start to sum the input signal pulses. This count continues for precisely one second, during which the 4-bit data latches prevent the counter outputs from reaching the display. At the end of the one-second timing period the GATE closes and terminates the count, and simultaneously an En pulse is fed to the set of data latches; this pulse has a width of (say) 100ns or greater, so all the counters have settled down and any glitches have disappeared by the time the En pulse terminates and latches the summed one-second end-count, which is fed to the display and reveals the input test signal's frequency, in Hz (cycles per second). The timing sequence is then complete, but a few moments later it starts to repeat, with the counters resetting and then counting the input pulses for another second, during which time the display gives a steady reading of the results of the previous count, and so on.

The *Figure 13.4* circuit thus generates a stable 'frequency' display that is regularly updated. In practice, the actual count period can be made any decade multiple or submultiple of one second, provided that the output display is suitably scaled. Thus, on a 3-digit display, a 'count' period of one second gives a maximum frequency reading of 999Hz, and a 1mS period gives a maximum frequency reading of 999kHz, and so on.

Data Latch ICs

The most popular and readily available TTL 4-bit transparent (level-triggered) data latch IC is the 74LS75, which has the functional diagram, etc., shown in *Figure 13.5*. Note the unusual positions of this IC's power

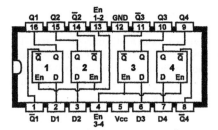

Figure 13.5 Functional diagram of the 74LS75 (or 74HC75) 4-bit level-triggered data latch IC

supply pins, and also note that the less readily available 74LS375 4-bit data latch IC is internally identical to the 74LS75, but has normal power supply pin allocations, as shown in *Figure 13.6*. Both of these ICs in fact house two independent 2-bit data latches, but can be made to function as standard 4-bit data latches of the *Figure 13.2*

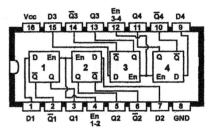

Figure 13.6 Functional diagram of the 74LS375 (or 74HC375) 4-bit level-triggered data latch IC

type (which are transparent when En is high but latch when En is low) by simply joining their En pins together; *Figures 13.7(a)* and *(b)* show the basic connections.

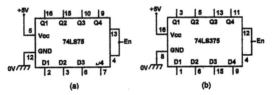

Figure 13.7   Way of using the *(a)* 74LS75 (or 74HC75) or *(b)* 74LS375 (or 74HC375) as a basic 4-bit data latch

Note that both of these ICs have a full set of Q and NOT-Q outputs, enabling any individual 4-bit code to be decoded via a 4-input AND (or NAND) gate. The BCD number '9', for example, has the DCBA binary code '1001', so can be decoded by simply ANDing the Q_D, NOT-Q_C, NOT-Q_B and Q_A outputs, which all go high when BCD '9' is present.

If you need an 8-bit level-triggered LS TTL data latch and don't need NOT-Q outputs, or do need 3-state outputs, the cheapest option is to use the 74LS373, which has the functional diagram shown in *Figure 13.8*. This 8-bit latch is transparent when the En terminal is high and latches when it is low; the outputs can be set into the high-impedance (3-state) mode by biasing pin-1 high (this pin must be biased low for normal operation).

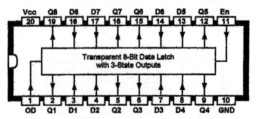

Figure 13.8   Functional diagram of the 74LS373 8-bit data latch IC

Several transparent data latch ICs are available in the '74HC'-series CMOS range, three of the most popular being the 74HC75 and 74HC375 4-bit types and the 74HC373 8-bit type. These are simply fast CMOS versions of the 74LS75, 74LS375 and 74LS373 respectively and have identical functional diagrams to those ICs.

Very few transparent data latch ICs are available in the '4000B'-series CMOS range. The best known example is the 4042B 4-bit data latch, which has the functional diagram and Truth Table shown in *Figure 13.9*. The four latches in this IC are controlled via a single ENABLE (En) terminal, and the device is unusual in that it provides the option, via its pin-6 POL (polarity) terminal, of either normal or inverted ENABLE action, as indicated in the Truth Table. Thus, if POL is set to logic-1, the latches are transparent when En is at logic-1 and latch the D data when En is at logic-0, but give the reverse of this action if POL is set to logic-0.

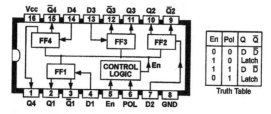

Figure 13.9 Functional diagram and Truth Table of the 4042B 4-bit 'transparent' data latch IC

Before leaving this subject, note that transparent data latches are most widely used as elements in 7-segment digital display driving systems of the types already shown in *Figures 13.3* and *13.4*, and are often incorporated in dedicated 'latch/decoder/7-segment-driver' ICs; two CMOS ICs of this type are described later in this chapter.

Shift Registers

In digital electronics, a 'register' is an element that provides temporary storage for one or more bits of binary data; a 'shift register' is one in which the stored bits can be displaced within the register – one step at a time – by applying suitable clock pulses. The simplest type of shift register is that shown in *Figure 13.10*, which is a synchronous 'bucket brigade' data shifter of the type already shown (in *Figure 10.40*) and described in Chapter 10. Here, the bit of binary data present at the input of FF1 is passed to FF1's output on the application of the first clock pulse, then to the output of FF2 on the second pulse, to the output of FF3 on the third pulse, and finally to the output of FF4 on the fourth pulse. The circuit can hold four bits of data at any given moment, and can store it indefinitely; when the data is needed, it can be clocked out (in serial form) by applying another set of clock pulses.

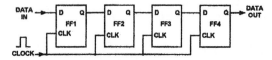

Figure 13.10   Basic 4-bit serial-in/serial-out (SISO) shift register circuit

The *Figure 13.10* circuit's data is clocked in and out in serial form, so it is known as a serial-in/serial-out or SISO shift register; it is useful for storing binary signals or delaying them by a fixed number of clock pulses, but not for much else. Its value can be greatly increased by converting it to a serial-in/parallel-out (SIPO) shift register by simply taking the 'parallel' outputs from the Q terminals of all four flip-flops, as shown in *Figure 13.11*; it is then useful for converting serial data into parallel form. This basic type of register can be made to give both serial and parallel outputs by adding a 'serial-out' connection as shown dotted in the diagram; such a unit is known as a SIPO/SISO shift register.

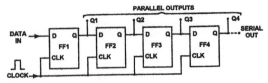

Figure 13.11   Basic 4-bit serial-in/parallel-out (SIPO) shift register circuit

The value of the basic shift register can be further increased by fitting it with PRESET terminals, as shown in *Figure 13.12*, so that it can be directly loaded with four bits of parallel data via pins P1 to P4. If the resulting circuit has no serial input facility, it is known as a parallel-in/serial-out (PISO) shift register, but if it does have a serial input facility (as shown dotted in the diagram) it is known as a PISO/SISO shift register. If it is configured to have only parallel-in and parallel-out facilities, it is known as a PIPO shift register. If a register has both parallel and serial inputs and parallel and serial outputs, it is simply called a 'universal' shift register.

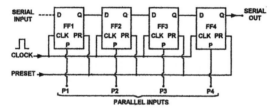

Figure 13.12   Basic 4-bit parallel-in/serial-out (PISO) shift register circuit

Figure 13.13 shows, in greatly simplified form, one widely used shift register application. Suppose here that a stack of 8-bit data words needs to be shifted from one point to another; normally, eight data links and one clock and one common link would be needed for this task, making a total of ten links, but by using shift registers the number of links can be reduced to three. This is achieved by first changing each 8-bit parallel word into serial form via a PISO shift register, then sending it and the clock and common signals down the 3-line link to be converted back into 8-bit parallel form

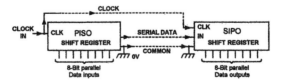

Figure 13.13   Shift registers used to transfer 8-bit parallel data via a 3-line connection

via a SIPO shift register at the destination point, as shown. This same basic technique can be used for transferring multi-bit words into or out of modern memory or data processing ICs in which only single IN, OUT, CLK and COMMON pins are allocated to these tasks.

The basic shift registers of *Figures 13.10* to *13.12* give an action in which the stored data shifts one step to the right on each application of a clock pulse, and are thus known as right-shift registers. But, just as an ordinary counter can be configured to count either up or down, a register can easily be configured to shift data to the right or left, or to have its direction selectable via a control terminal. A large variety of shift register configurations are thus possible.

In practice, most modern '74LS'-series shift register ICs are of the 'universal' type, one of the few exceptions being the 74LS164 8-bit SIPO shift register. Of the universal types, the 74LS195 is a plain 4-bit IC, the 74LS295 and 74LS395 are 4-bit types with 3-state outputs, the 74LS95 and 74LS194 are 4-bit right/left shift types, and the 74LS299 and 74LS323 are 8-bit types with 3-state outputs.

Among popular '74HC'-series types, the 74HC91 is an 8-stage SISO type, the 74HC164 is an 8-bit SIPO/SISO type, and the 74HC194 is a 4-bit bidirectional 'universal' type with CLEAR.

The oldest current-production '4000B'-series shift register IC is the 4006B, which the manufacturers simply describe as an '18-stage static shift register'. *Figure 13.14* shows the functional diagram of this IC, which in

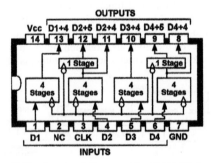

Figure 13.14 Functional diagram of the 4006B 18-stage static shift register IC

fact houses four separate 4-stage shift registers, two of which are followed by an extra stage, with all stages controlled by a common clock signal that shifts data on its falling edges. The IC can be used as either a 4-bit PIPO type or (by cascading stages) as a SISO type with 4, 5, 8, 9, 10, 12, 13, 14, 16, 17, or 18 effective stages. (Note: On most versions of this IC, pin-2 is not internally connected, but on the RCA version it gives an extra D1 + 4 output that is delayed one-half clock cycle.)

Among other popular '4000B'-series types, the 4014B is an 8-stage PISO/SISO type, the 4015B is a Dual 4-stage SIPO type with RESET, the 4021B is an 8-stage PISO/SISO type, the 4031B is a 64-stage SISO type, the 4035B is a 4-bit PIPO type, and the 4094B is an 8-stage SIPO type with buffered 3-state outputs.

Logic Comparators

A digital 'logic' or 'magnitude' comparator IC is one that compares the binary codes of two words (A and B) of the same bit size and has outputs that indicate whether code A is greater than, smaller than, or equal to code B. This action is useful in, for example, triggering some action when a counting chain reaches a certain value, etc. The best known ICs of this type are the 74LS85, 74HC85 and 4063B 4-bit comparators (see *Figure 13.15*), which are functionally identical and which can be cascaded with

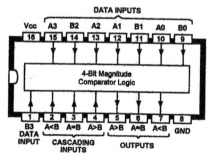

Figure 13.15 Functional diagram of the 74LS85, 74HC85, or 4063B 4-bit magnitude comparator IC

others of their type to make a comparator of any desired bit size. Note that, as well as having four input terminals for each 4-bit word, plus three active-high output terminals (notated A > B, A = B, and A < B), this IC also has three input terminals (also notated A > B, A = B, and A < B) that are used to implement cascaded operations.

The 74LS/HC85 and 4063B are very easy to use, as shown in *Figure 13.16*. If only 4-bit words are being compared, a single IC is used, connected in the manner of IC1, with the A > B and A < B cascading terminals grounded and the A = B cascading terminal biased high, and with the three outputs taken from pins 5, 6 and 7. If the words have bit lengths that are whole-number multiples of four, they can be compared by cascading an appropriate number of 74LS/HC85 or 4063B on a basis of 4-bits per IC (e.g. three ICs for 12-bit comparison, etc.); in this case the four least-significant bits (LSBs) must be allocated to IC1, which must be connected in the way already described, and the four most-significant bits (MSBs) must be allocated to the final IC (IC*N*), and so on; in this case, the three outputs of each lower-order IC must be connected to the three 'cascading' inputs of the following IC, the final outputs being taken from the last IC in the chain.

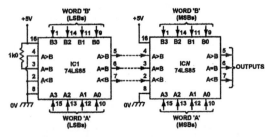

Figure 13.16 Basic way of using the 74LS85 (or 74HC85 or 4063B) 'comparator' ICs

Code Converters

Code converter ICs are widely used in digital electronics to change electronic codes from one format into another.

Some types (such as the 4514B and 4515B) are full 4-bit binary decoders, with an individual output for each of the 16 possible code numbers. Others (such as the 4028B) are 4-bit BCD-to-decimal decoders, with ten individual outputs. The most widely used type of converter is the BCD to 7-segment decoder/driver IC, which takes the 4-bit BCD output of a counter, etc., and converts it into a form suitable for directly driving a 7-segment LED or LCD digital display; some ICs of this type incorporate a 4-bit data latch that enables the display to be 'frozen'. Before looking at specific ICs of these types, however, it is first necessary to note a few points about 7-segment digital display basics, as follows:

A 7-segment display is a unit that houses seven independently accessible photoelectric elements such as LEDs or liquid crystal segments, arranged in the form shown in *Figure 13.17*. The segments are conventionally notated from *a* to *g* in the manner shown, and they can

Figure 13.17   Standard form and notations of a 7-segment display

be made to display any number from 0 to 9 or letter from A to F (in a mixture of upper and lower case) by activating these segments in various combinations, as shown in the Truth Table of *Figure 13.18*.

SEGMENTS (✓ = ON)							DISPLAY	SEGMENTS (✓ = ON)							DISPLAY
a	b	c	d	e	f	g		a	b	c	d	e	f	g	
✓	✓	✓	✓	✓	✓		0	✓	✓	✓	✓	✓	✓	✓	8
	✓	✓					1	✓	✓	✓			✓	✓	9
✓	✓		✓	✓		✓	2	✓	✓	✓		✓	✓	✓	A
✓	✓	✓	✓			✓	3			✓	✓	✓	✓	✓	b
	✓	✓			✓	✓	4	✓			✓	✓	✓		C
✓		✓	✓		✓	✓	5		✓	✓	✓	✓		✓	d
✓		✓	✓	✓	✓	✓	6	✓			✓	✓	✓	✓	E
✓	✓	✓					7	✓				✓	✓	✓	F

Figure 13.18   Truth Table of a 7-segment display

Practical 7-segment displays need at least eight external connectors, one of which acts as the 'common' terminal. If the display is an LED type, the seven individual LEDs may be arranged as shown in *Figure 13.19*, with all LED

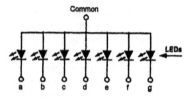

Figure 13.19   Schematic diagram of a common-anode 7-segment LED display

anodes connected to the common terminal, or as in *Figure 13.20*, with all LED cathodes connected to the common terminal; in the former case the 7-segment display unit is known as a common-anode type, and in the latter it is called a common-cathode type.

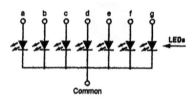

Figure 13.20   Schematic diagram of a common-cathode 7-segment LED display

In most practical applications, 7-segment displays are driven – via a suitable decoder/driver IC – from a 4-bit BCD input, and the IC and display are connected as shown in *Figure 13.21*. The IC houses a moderately

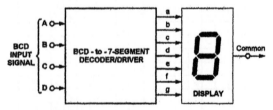

Figure 13.21   Basic connections of a BCD to 7-segment decoder/driver IC

complex set of logic gates, as is implied by *Figure 13.22*, which shows the standard relationship between the BCD input codes and the displayed 7-segment numerals. In practice, BCD to 7-segment decoder/driver ICs are usually available in a dedicated form suitable for

BCD Signal				DISPLAY	BCD Signal				DISPLAY
D	C	B	A		D	C	B	A	
0	0	0	0	0	0	1	0	1	5
0	0	0	1	1	0	1	1	0	6
0	0	1	0	2	0	1	1	1	7
0	0	1	1	3	1	0	0	0	8
0	1	0	0	4	1	0	0	1	9

0 = logic low
1 = logic high

Figure 13.22   Truth Table of a BCD to 7-segment decoder/driver

driving only a single type of display unit, e.g. a common-anode LED type, common-cathode LED type, or a liquid-crystal display (LCD). *Figures 13.23* to *13.25* show the basic ways of interconnecting each of these IC and display types.

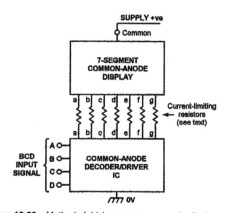

Figure 13.23   Method of driving a common-anode display

Note in the LED circuits that in most cases a current-limiting resistor (about 150R with a 5V supply, or 680R at 15V) must be wired in series with each display segment. To drive a common-anode display (*Figure*

13.23), the driver must have an active-low output, in which each segment-driving output is normally high, but goes low to turn a segment on. To drive a common-cathode display (*Figure 13.24*), the driver must have an active-high output.

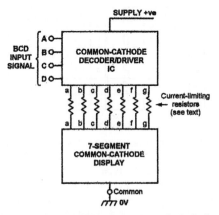

Figure 13.24   Method of driving a common-cathode display

In the *Figure 13.25* LCD-driving circuit the display's common BP (back-plane) terminal and the IC's PHASE input terminal must be driven by a symmetrical squarewave (typically 30Hz to 200Hz) that switches fully between the two supply rail voltages (0V and V+), as shown. The full explanation for this is a little complicated, as follows.

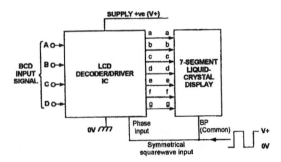

Figure 13.25   Method of driving a liquid-crystal display (LCD)

To drive an LCD segment, the driving voltage must be applied between the segment and BP terminals. When the voltage is zero, the segment is effectively invisible. When the drive voltage has a significant positive or negative value, however, the segment becomes effectively visible, *but* if the drive voltage is sustained for more than a few hundred milliseconds the segment may become permanently visible and be of no further value. The way around this problem is, in principle, to drive the segment on via a perfectly symmetrical squarewave that switches alternately between identical positive and negative voltages, and thus has zero DC components and will not damage the LCD segment even if sustained permanently. In practice, this type of waveform is actually generated with the aid of an EX-OR True/Complement generator (see Chapter 6, *Figure 6.97)*, connected as shown in *Figure 13.26(a)*.

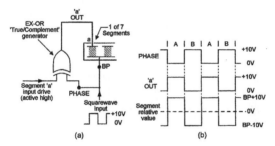

Figure 13.26   Basic LCD segment-drive circuit *(a)*, and voltage-doubling 'bridge-driven' segment waveforms *(b)*

In *Figure 13.26(a)*, the basic segment 'a' input drive (which is active-high) is connected to one input of the EX-OR element, and the other EX-OR input terminal (which is notated PHASE) is driven by a symmetrical squarewave that switches fully between the circuit's supply rail voltages (shown as 0V and +10V) and is also applied to the LCD display's BP pin. When the segment 'a' input drive is low, the EX-OR element gives a non-inverted (in-phase) 'a' output when the squarewave is at logic-0, and an inverted (anti-phase) 'a' output when the squarewave is at logic-1, and thus produces zero voltage difference between the 'a' segment and BP points under both these conditions; the segment is thus turned off under these conditions. When the segment 'a' input drive is high, the EX-OR element gives the same phase action as just described, but in this case the 'a' OUT pin is high and BP is low when the squarewave is at logic-0, and 'a' OUT is low and BP is high when the squarewave is at logic-1; the segment is thus turned on under these conditions.

Figure 13.26(b) shows the circuit waveforms that occur when the 'a' segment is turned on, with the 'a' segment and BP driven by anti-phase squarewaves. Thus, in part A of the waveform the segment is 10V positive to BP, and in part B it is 10V negative to BP, so the LCD is effectively driven by a squarewave with a peak-to-peak value of 20V but with zero DC value. This form of drive is generally known as a voltage-doubling 'bridge drive' system. In practice, many LCD-driving ICs (such as the 4543B) incorporate this type of drive system in the form of a 7-section EX-OR gate array interposed in series with the segment output pins, with access to its common line via a single PHASE terminal.

Note that any 'active-high' 7-segment LED-driving Decoder IC can be used to drive a 7-segment LCD display by interposing a 'bridge driven' 7-section EX-OR array (etc.) between its segment output pins and the segment pins of the LCD display, as shown in *Figure 13.27*, which shows a 7448 or 74LS48 TTL IC circuit so adapted.

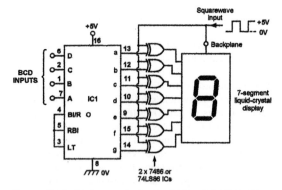

Figure 13.27 Basic way of using the 7448 or 74LS48 to drive a liquid-crystal display

Multi-digit Displays

In reality, 7-segment displays are usually used in multi-digit applications such as that shown in *Figure 13.28*, which shows the basic elements of a 4-digit counter/display circuit that can give a maximum reading of '9999'. Note that if this circuit is used to measure a count of (say) 27 it will actually give a reading of '0027', unless steps are taken to automatically suppress the two (unwanted) leading zeros. Similarly, if the same display is used on a 4-digit voltmeter scaled to read a maximum of 9.999 volts, it will give a reading of 0.100 volts if fed with a 0.1V input, unless steps are taken to suppress the two trailing zeros.

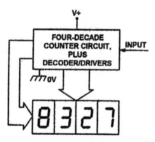

Figure 13.28 Basic elements of a 4-digit counter/display circuit

In practice, many modern decoder/driver ICs have facilities for giving automatic blanking of leading and/or trailing zeros, using the basic ripple-blanking techniques illustrated in *Figures 13.29* and *13.30*. Note in these diagrams that each decoder/driver IC is provided with ripple-blanking input (RBI) and output (RBO) terminals; if these terminals are active-high their actions are such that the IC gives normal decoder/driver action and RBO is disabled (driven low) when RBI is biased low, but the display is blanked and RBO is driven high under the 'zero' (BCD input '0000') condition when RBI is biased high. Thus, the RBO terminal is normally low and goes high only if a BCD '0000' input is present at the same time as the RBI terminal is high. With these facts in mind, refer now to *Figures 13.29* and *13.30*.

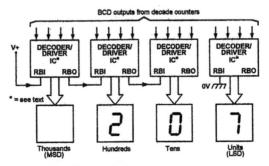

Figure 13.29 Ripple blanking used to give leading-zero suppression in a 4-digit counter

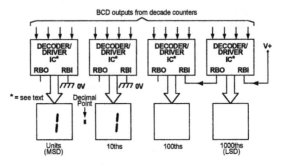

Figure 13.30 Ripple blanking used to give trailing-zero suppression in a 4-digit voltmeter

Figure 13.29 shows the ripple-blanking technique used to provide leading-zero suppression in a 4-digit display that is reading a count of 207. Here, the RBI input of the 'thousands' or MSD decoder/driver IC is tied high, so this display is automatically blanked and RBO is driven high in the presence of a zero. Consequently, RBI of the 'hundreds' IC is driven high under this condition; its display reads '2' and its RBO terminal is thus low. The RBI input of the 'tens' unit is thus also low, so its display reads '0' and its RBO output is low. The least significant digit (LSD) is that of the 'units' readout and does not require zero suppression, so its RBI input is grounded and it reads '7'. The display thus gives an overall reading of '207'.

Note in the *Figure 13.29* leading-zero suppression circuit that ripple blanking feedback is applied backwards, from the MSD to the LSD. *Figure 13.30* shows how trailing zero suppression can be obtained by reversing the direction of feedback, from the LSD to the MSD. Thus, when an input of 1.1V is fed to this circuit the LSD is blanked, since its BCD input is '0000' and its RBI input is high. Its RBO terminal is high under this condition, so the '100ths' digit is also blanked in the presence of a '0000' input.

Not all decoder/driver ICs are provided with RBI and RBO ripple-blanking terminals, and some of those that are have an active-low action. If a decoder/driver IC does not incorporate a ripple-blanking feature, it can usually be obtained by adding external logic similar to that shown in *Figure 13.31*, with the RBO terminal connected to the *blanking* input pin of the decoder/driver IC. In *Figure 13.31* (an active-high circuit), the output of the 4-input NOR gate goes high only in the presence of a '0000' BCD input, and the RBO output goes high only if this input is present while RBI is high.

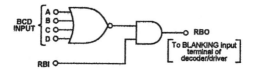

Figure 13.31   DIY ripple-blanking logic (active-high type)

TTL Decoder/driver ICs

In TTL, the two most popular decoder/driver ICs are the old 7447A and 7448 types and their modern counterparts, the 74LS47 and 74LS48. These ICs are functionally very similar, as can be seen from *Figures 13.32(a)* and *(b)*; all four ICs have integral ripple-blanking facilities. The 7447A/74LS47 has an active-low output and is specifically designed for driving a common-anode LED display via external current-limiting resistors R_X (typically 150R), as shown in

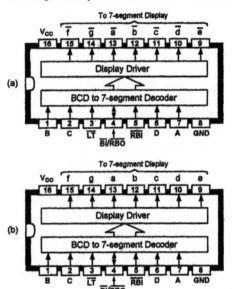

Figure 13.32 Functional diagrams of the *(a)* 7447A/74LS47 and *(b)* 7448/74LS48 BCD to 7-segment decoder/driver ICs

Figure 13.33. The 7448/74LS48 has an active-high output designed for driving a common-cathode LED display in a manner similar to that of *Figure 13.33*, but

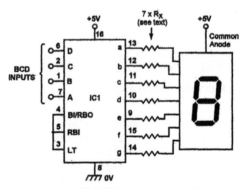

Figure 13.33 Basic ways of using a 7447A or 74LS47 to drive a common-anode LED display

with the display's common terminal taken to ground. In all cases, the R_X current-limiting resistors must be chosen to hold the segment currents below the following absolute limits:

4774A = 40mA;   74LS47 = 24mA;
7448/74LS48 = 6mA.

Note that *Figure 13.27* has already shown how the 7448/74LS48 can be used to drive a liquid-crystal display (LCD), using a pair of 7486 or 74LS86 Quad 2-input EX-OR gate ICs and an external 50Hz squarewave to apply the necessary phase drive signals to the display, as already described.

Note from *Figure 13.32* that each of these ICs has three active-low input 'control' terminals, these being designated NOT-LT (Lamp Test), NOT-BI/NOT-RBO, and NOT-RBI. The NOT-LT terminal drives all display terminals on when it is driven low when the NOT-BI/NOT-RBO terminal is open-circuit or high. When the NOT-BI/NOT-RBO terminal is pulled low all outputs are blanked; this terminal also functions as a ripple-blanking output terminal. *Figure 13.34* shows how to connect the ripple-blanking terminals to give leading-zero suppression on the first three digits of a 4-digit display (this basic circuit can be used with any IC within the 7447/7448 family of devices).

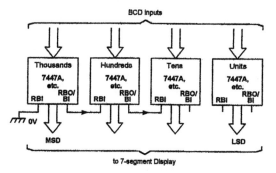

Figure 13.34   Method of applying leading-zero suppression to the first three digits of a 4-digit display, using 7447A or 74LS47 ICs

CMOS Decoder/driver ICs

In '4000B'-series CMOS, one of the most popular 'decoder' ICs is the 4028B BCD-to-decimal decoder. *Figure 13.35* shows the functional diagram of the 4028B, which has four input terminals (notated A to D) and ten output terminals (notated 0 to 9) and simply gives direct decoding of the ten BCD or 4-bit binary numbers 0 to 9 inclusive; input terminals A to D correspond to Q1 to Q4 of these codes. Note that this IC can be used as a 3-bit octal decoder by grounding the 'D' terminal, applying the 3-bit input to the A-to-C terminals, and taking the eight outputs from the 0 to 7 pins.

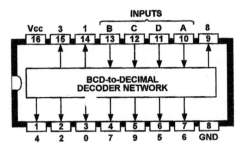

Figure 13.35   Functional diagram of the 4028B BCD-to-decimal decoder

The most popular full 4-bit decoder ICs, which provide an individual output for each of 16 possible code numbers, are the 4514B and 4515B, which are also available in the '74HC'-series as the 74HC4514 and 74HC4515. *Figures 13.36* and *13.37* show the functional diagrams of these ICs, which each incorporate a 4-bit latch that can be controlled via the pin-1 F/NOT-L terminal, and have a master INH (inhibit) control on pin-23. The major difference between the two basic IC types is that the 4514B and 74HC4514 have active-high outputs, in which all except the selected output are normally low, and the 4515B and 74HC4515 have active-low outputs, in which all but the selected output are normally high.

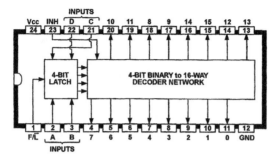

Figure 13.36 Functional diagram of the 4514B or 74HC4514 4-bit to 16-way latched decoder with active-high outputs

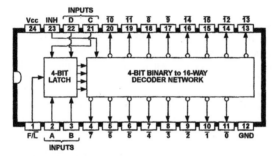

Figure 13.37 Functional diagram of the 4515B or 74HC4515 4-bit to 16-way latched decoder with active-low outputs

In these ICs, the INH control is normally held low, and when driven high disables all decoding functions; it drives all outputs low in the 4514B type, or high in the 4515B type, irrespective of the states of all other pins. Each IC acts as a direct decoder when the F/NOT-L terminal is high, but when the terminal is pulled low it latches the prevailing input code and holds it for as long as the terminal remains low; the latched code is decoded and fed to the output in the normal way.

The most popular '4000B'-series BCD to 7-segment LED-driving IC is the 4511B (also available as the 74HC4511), which has a built-in data latch and has the functional diagram shown in *Figure 13.38*. This IC does not incorporate ripple-blanking facilities, but is otherwise ideally suited to driving common-cathode LED displays, since its outputs can each source up to

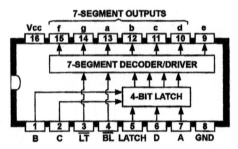

Figure 13.38   Functional diagram of the 4511B or 74HC4511 7-segment latch/decoder/LED-driver IC

25mA. The 4511B is very easy to use, and has only three input control terminals; of these, the NOT-LT (pin-3) pin is normally tied high, but turns on all seven segments of the display when pulled low. The NOT-BL (pin-4) terminal is also normally tied high, but blanks (turns off) all seven segments when pulled low. Finally, the LATCH (pin-5) terminal enables the IC to give either direct or latched decoding operation; when LATCH is low, the IC gives direct decoding operation, but when LATCH is taken high it holds the display in a 'frozen' state.

Finally, the most popular '4000B'-series BCD to 7-segment LCD-driving IC is the 4543B (also available as the 74HC4543), which also has a built-in data latch and has the functional diagram shown in *Figure 13.39*. This IC incorporates an EX-OR array (of the type shown in *Figures 13.26* and *13.27*) in its output driver network, which can source or sink several mA of output current.

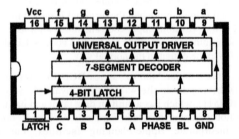

Figure 13.39   Functional diagram of the 4543B or 74HC4543 'universal' 7-segment latch/decoder/driver IC

This feature enables the IC to act as a 'universal' unit that can drive common-cathode or common-anode LED or liquid-crystal 7-segment displays with equal ease, as shown in *Figures 13.40* and *13.41*.

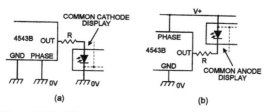

Figure 13.40 Way of using the 4543B (or 74HC4543) to drive (a) common-cathode or (b) common-anode 7-segment LED displays

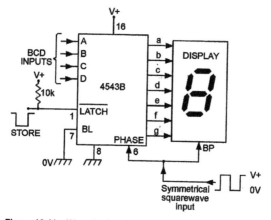

Figure 13.41 Way of using the 4543B (or 74HC4543) to drive a 7-segment LCD

The 4543B (or 74HC4543) is delightfully easy to use. Its BL (blank) terminal is normally grounded, and blanks the entire display when pulled high, and the NOT-LATCH (pin-1) terminal gives direct decoding when biased high or latched decoding (e.g. a 'frozen' display) when pulled low. The IC can be used to drive a common-cathode LED display by connecting it as in *Figure 13.40(a)*, with the PHASE control grounded, or a common-anode LED display by connecting it as in

Figure 13.40(b), with the PHASE control wired to the V+ line; in both cases, resistor 'R' acts as a segment current-limiting resistor. The IC can be used to drive a liquid-crystal display (LCD) by connecting it as shown in *Figure 13.41* and driving the IC's PHASE and the display's BP pins via a symmetrical squarewave that switches fully between the circuit's supply-rail values.

14 Special-purpose ICs and Circuits

So far, this Volume has dealt with mainly run-of-the mill digital ICs such as logic gates, bilateral switches, flip-flops, counters, latches, and other devices of the type that the average professional digital circuit designer is likely to use on a regular basis. But there are other useful types of digital IC that may be needed only very rarely, and among them are dedicated multiplexers, decoders and demultiplexers, addressable latches, full-adders, and bus transceivers. This chapter explains how to use these types of special-purpose ICs, and concludes by giving functional descriptions of a few other unusual and very rarely used types of logic IC.

Multiplexing Basics

Multiplexers and demultiplexers were briefly defined in Chapter 7, where it was explained that some CMOS bilateral switch ICs are designed for use as versatile multi-channel multiplexer/demultiplexer units. Several '74LS'-series TTL ICs and a few '74HC'-series CMOS types are, however, designed purely as dedicated multiplexers or as dedicated demultiplexer devices, and some of these ICs are described in the next two sections of this chapter.

A multiplexer is a device that enables two or more signals to be selected and combined into a single output that can subsequently be demultiplexed in a way that enables the original signals to be retrieved. *Figure 14.1* illustrates the basic principle of a multiplexing system. Imagine here that the two switches are motor driven and continuously rotate and are somehow remotely ganged so that SW2 is in position '1' when SW1 is in position '1', and so on; consequently, the 4-bit data from the four input lines is repeatedly sequentially inspected via SW1 and converted into serial form (multiplexed), and then shoved down the data link to SW2, where it is

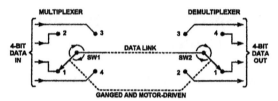

Figure 14.1 Basic principles of a multiplexed data transfer system

demultiplexed and reappears in its original 4-bit form on four separate output lines. The big feature of this system is, of course, that it enables a whole stack of parallel data to be transmitted – in serial form – via a single data link such as an electric or fibre-optic cable or a wireless carrier wave, etc.

Figure 14.2 shows, in greatly simplified form, an electronic version of the above circuit. Here, at the transmitter end of the system, SW1 is replaced by a 4-input multiplexer, the action of which is such that any of the four data input lines can be coupled to the OUT line by applying a suitable 2-bit binary address code (00, 01, 10, or 11). These codes are generated sequentially by the divide-by-4 counter (consisting of two cascaded divide-by-2 flip-flops), which is driven by a clock-pulse generator that produces narrow trigger pulses. The outputs of the multiplexer and the clock-pulse generator are mixed together in the 2-input AND gate and transmitted down the single data link; at the receiver end of the system, the clock pulses are extracted from the data link and used to drive another divide-by-4 counter that generates address codes for the 1-line to 4-line demultiplexer, which reconstructs the original input data and puts it out on four separate lines.

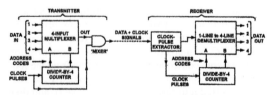

Figure 14.2 Basic circuit of a multiplexed data transfer system

The above circuit is, of course, greatly simplified, and in practice would need the addition of a pulse synchronization system and a few other refinements to make it work properly, but it does serve to illustrate the basic multiplexing principle. Note in particular that multiplexers and demultiplexers are really meant to form individual elements in a highly specialized type of system, but that in practice a multiplexer actually functions as an addressable data selector (like SW1 in *Figure 14.1*), and a demultiplexer functions as an addressable data distributor (like SW2 in *Figure 14.1*) or as a binary-code 'decoder', and in these modes both types of device are so useful that they are usually described as 'multiplexer/data-selector' and 'demultiplexer/decoder' ICs.

Note in *Figure 14.1* that SW1 and SW2 can both pass signals in either direction, and can thus be used as either multiplexers or demultiplexers by simply placing them in the appropriate part of the system. In practice, however, dedicated multiplexers and demultiplexers can only pass signals in a single direction.

Popular Multiplexer ICs

The three best known types of current-production TTL multiplexer IC are the 74LS157 Quad 2-input IC, the 74LS153 Dual 4-input IC, and the 74LS151 8-input IC, which are also available in similarly numbered '74HC'-series fast CMOS versions. *Figure 14.3* shows the

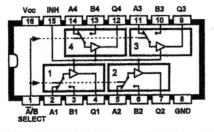

Figure 14.3 Functional diagram of the 74LS157 (or 47HC157) Quad 2-input data-selector/multiplexer IC

functional diagram of the 74LS157 (or 74HC157), which effectively houses four ganged 2-way ('A' or 'B') switches with buffered outputs that can be disabled (driven low) by biasing INH pin-15 high; the switch positions can be selected via pin-1, which selects position 'A' when biased low or 'B' when biased high. The 74LS157 is very easy to use, and *Figure 14.4* shows how to connect it as a data-selector that can select either of two 4-bit input words via pin-1. This circuit is useful in applications where, for example, either of two 4-bit codes needs to be sent to the PRESET terminal of a counter/divider IC, etc.

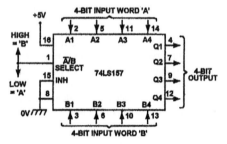

Figure 14.4   Normal connections for using the 74LS157 (or 74HC157) as a 4-bit data-word selector

Figure 14.5 shows the functional diagram of the 74LS153 (or 74HC153), which effectively houses two ganged 4-way (A to D) switches with buffered outputs that can be disabled (driven low) by biasing the appropriate INH terminal (pins 1 or 15) high; the switch positions can be selected by applying the appropriate BA binary codes to pins 2 and 14, as indicated in the

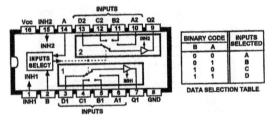

Figure 14.5   Functional diagram and Data Selection Table of the 74LS153 (or 74HC153) Dual 4-input data selector/multiplexer IC

Figure 14.5 table. *Figure 14.6* shows how to connect the IC as a Dual 4-way input selector, in which each 'switch' outputs the input data that is selected via the BA 'select' code. Note that the IC can be used as a Dual 2-bit decoder by applying the 2-bit code to the BA terminals and tying all but one of each switch's four inputs low, so that the switch's output goes high only when the desired 2-bit code is present, as indicated in the IC's Data Selection Table; thus, the '10' BA code can be detected by tying only the 'C' input high, etc.

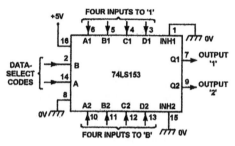

Figure 14.6 Normal connections for using the 74LS153 (or 74HC153) as a Dual 4-way input selector

Figure 14.7 shows the functional diagram of the 74LS151 (or 74HC151), which effectively houses a single 8-way ('A1' to 'H1') switch with a buffered output that can be disabled (driven low) by biasing the INH terminal high; the switch positions can be selected by applying the appropriate CBA 3-bit binary codes to pins 9–10–11, as indicated in the *Figure 14.7* table. *Figure 14.8* shows how to connect the IC as an 8-way input selector, which outputs whichever switch input is

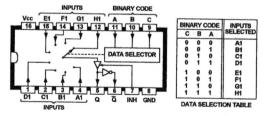

Figure 14.7 Functional diagram and Data Selection Table of the 74LS151 (or 74HC151) 8-input data selector/multiplexer IC

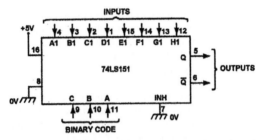

Figure 14.8 Normal connections for using the 74LS151 (or 74HC151) as an 8-way input selector

selected via the CBA 'select' code; a direct output is available on pin-5, and an inverted one on pin-6. Alternatively, *Figure 14.9* shows the IC wired as a 3-bit decoder, with the 3-bit code applied to the CBA terminals and with all but one of the eight inputs tied low, so that the pin-5 output goes high only when the desired 3-bit code is present (see the *Figure 14.7* Data Selection Table); the IC is shown connected to detect the '101' code, which selects the F1 input, which in this case is tied high.

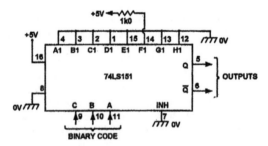

Figure 14.9 Normal connections for using the 74LS151 (or 74HC151) as a 3-bit decoder (shown set for '101' decoding)

Popular Demultiplexer ICs

The three best known types of current-production TTL decoder/demultiplexer ICs are the 74LS139 Dual 4-output IC, the 74LS138 8-output IC, and the 74LS154

16-output IC, which all have active-low outputs and are also available in similarly numbered '74HC'-series versions. *Figure 14.10* shows the functional diagram of

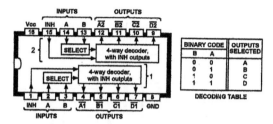

Figure 14.10 Functional diagram and Decoding Table of the 74LS139 (or 74HC139) Dual 4-way decoder/demultiplexer IC

the 74LS139 (or 74HC139), which effectively houses two independent 4-way (NOT-A to NOT-D) selectors with active-low buffered outputs that can all be disabled (driven high) by biasing the INH terminal high; the switch positions are selected via a 2-bit BA binary code, as shown in the diagram's Decoding Table. The 74LS139 can be used as a Dual 2-bit 4-way (2-line to 4-line) decoder by connecting it as shown in *Figure 14.11*,

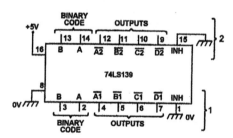

Figure 14.11 Normal connections for using the 74LS139 (or 74HC139) as a Dual 2-line to 4-line decoder

with the INH terminals grounded and the decoded outputs taken from the appropriate output terminals (see the IC's Decoding Table); thus, the NOT-C output is normally high, and goes low only in the presence of a '10' BA input code, etc. The IC can be used as a demultiplexer by using the INH terminal as the data input.

Figure 14.12 shows the functional diagram and Decoding Table of the 74LS138 (or 74HC138), which effectively houses an 8-way selector with active-low buffered outputs that can all be disabled (driven high) by biasing the I1 or I2 INH terminal high or the NOT-I3 terminal low; the switch positions are selected via a 3-bit CBA binary code, as shown in the diagram's Decoding Table. The 74LS138 can be used as a 3-bit 8-

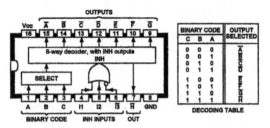

Figure 14.12   Functional diagram and Decoding Table of the 74LS138 (or 74HC138) 8-way decoder/demultiplexer IC

way (3-line to 8-line) decoder by connecting it as shown in *Figure 14.13*, with the decoded outputs taken from the appropriate output terminals (see the IC's Decoding Table); each output is normally high, but goes low when activated by its 3-bit 'select' code. This IC can be used as a demultiplexer by using one of the active-low INH terminals as the data input, with the other two INH inputs disabled.

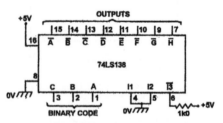

Figure 14.13   Normal connections for using the 74LS138 (or 74HC138) as a 3-line to 8-line decoder

Finally, *Figure 14.14* shows the functional diagram and Decoding Table of the 24-pin 74LS154 (or 74HC154), which effectively houses a 16-way selector with active-low buffered outputs that can all be disabled (driven

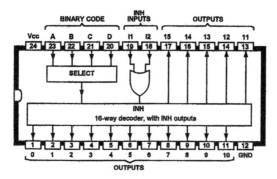

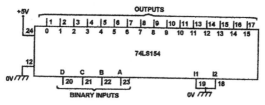

Figure 14.14 Functional diagram and Decoding Table of the 74LS154 (or 74HC154) 16-way decoder/demultiplexer IC

high) by biasing either of the two INH terminals high; the switch positions are selected via a 4-bit DCBA binary code, as shown in the Decoding Table. The 74LS154 can be used as a 4-bit 16-way (4-line to 16-line) decoder by connecting it as in *Figure 14.15*, with the decoded outputs taken from the appropriate output

Figure 14.15 Normal connections for using the 74LS154 (or 74HC154) as a 4-line to 16-line decoder

terminals (see the IC's Decoding Table); each output is normally high, but goes low when activated by its 4-bit 'select' code. The IC can be used as a demultiplexer by tying one INH input low and using the other INH terminal as a data input.

Addressable Latches

Chapter 13 gives a reasonably full description of conventional data latch ICs, in which groups of latch elements are ganged together and activated by a single clock signal. An 'addressable latch' IC, on the other hand, is a special unit in which a single clock signal and data input, etc., can be 'addressed' (applied) to individual latch elements among a group, one at a time; the 'address' takes the form of a binary code (2-bits for four latches, 3-bits for eight latches). The best known TTL IC of this type is the 74LS259 8-bit addressable latch, which is also available in a fast CMOS version as the 74HC259; *Figure 14.16* shows the functional diagram, Address Table, and Function Table of the 74LS259 (or 74HC259). This IC has four basic operating modes (see the Function Table), as follows:

1. When NOT-CLR is low and CLK is high, all latches are cleared and the Q0 to Q7 outputs go low.

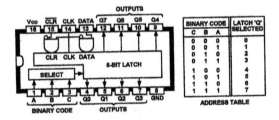

Figure 14.16 Functional diagram, Address Table, and Function Table of the 74LS259 (or 74HC259) 8-bit addressable latch IC

2. When NOT-CLR and CLK are low, the IC acts as a 3-line to 8-line decoder (or as a demultiplexer) in which the selected output is active-high and all other outputs are low.

3. When NOT-CLR is high and CLK is low, the IC acts as an addressable transparent latch, in which the Q output of the selected latch follows the data input, and all other latches remain in their previous states.

4. When NOT-CLR and CLK are high, the current data is latched into the selected latch.

The 74LS/HC259 is thus a reasonably versatile unit that can be used as an 8-bit serial-in to parallel-out converter, as a general-purpose memory unit, or as a decoder or demultiplexer, etc. *Figure 14.17* shows it used as a 3-line to 8-line decoder that drives the selected output high in the presence of the correct address code, and *Figure 14.18* shows it used as an 8-bit demultiplexer, with the serial input applied to the data terminal and the clock signal (which would also drive the 3-stage counter that generates the CBA binary 'select' code) applied to the CLK terminal.

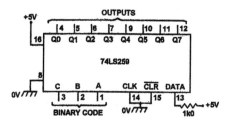

Figure 14.17 Connections for using the 74LS259 (or 74HC259) as a 3-line to 8-line decoder with active-high outputs

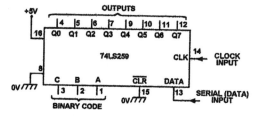

Figure 14.18 Connections for using the 74LS259 (or 74HC259) as an 8-bit demultiplexer

Figure 14.19 shows the functional diagram, Address Table, and Function Table of a popular CMOS 8-bit addressable latch IC, the 4099B. In this IC, the DIS (select DISABLE) input acts a bit like a clock terminal, and must be pulled low (to logic-0) to load the binary select code into the latches. When DIS is pulled low, the IC acts as an addressable latch when CLR is low, or as a demultiplexer when CLR is high. When DIS is biased high, the IC acts as an 8-bit memory when CLR is low, and all latches are reset (with their 'Q' outputs set low) when CLR is high.

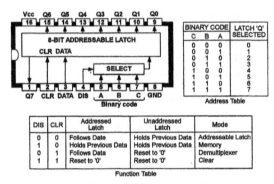

BINARY CODE			LATCH 'Q' SELECTED
C	B	A	
0	0	0	0
0	0	1	1
0	1	0	2
0	1	1	3
1	0	0	4
1	0	1	5
1	1	0	6
1	1	1	7

Address Table

DIS	CLR	Addressed Latch	Unaddressed Latch	Mode
0	0	Follows Date	Holds Previous Data	Addressable Latch
1	0	Holds Previous Data	Holds Previous Data	Memory
0	1	Follows Data	Reset to '0'	Demultiplexer
1	1	Reset to '0'	Reset to '0'	Clear

Function Table

Figure 14.19   Functional diagram, Address Table, and Function Table of the 4099B 8-bit addressable latch IC

Full-adder ICs

The basic principles of binary half-adder and full-adder circuits are outlined in Chapter 6. The two best known TTL ICs of the latter type are the 74LS83 and the 74LS283 4-bit full-adders, which are functionally identical but have different pin-outs, as shown in *Figures 14.20* and *14.21*. Each of these ICs generates a 4-bit plus CARRY output equal to the sum ('S') of two 4-bit (DCBA) input words ('1' and '2'); numbers of these ICs can be coupled together to carry out binary addition on words of any desired bit size.

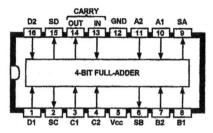

Figure 14.20   Functional diagram of the 74LS83 4-bit full-adder IC

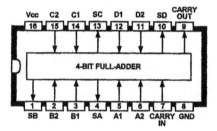

Figure 14.21   Functional diagram of the 74LS283 4-bit full-adder IC

Figure 14.22 shows the basic way of using the 74LS83 or 74LS283 as a single 4-bit adder; the CARRY IN terminal is tied low, word '1' is applied to one set of input terminals and word '2' is applied to the other, and the result of the addition appears on the 'S' output terminals. If the result of the addition is greater than decimal 15 (binary 1111), the CARRY OUT terminal goes to logic-1, and thus acts as a bit-5 output.

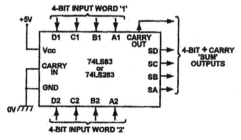

Figure 14.22   Normal circuit for using the 74LS83 or 74LS283 as a 4-bit full-adder

Figure 14.23 shows how to interconnect two 4-bit full-adders to make an 8-bit adder. The least-significant four bits of each input word are applied to IC1, which has its CARRY IN terminal tied low, and the four most-significant bits are applied to IC2, which has its CARRY IN terminal tied to the CARRY OUT of IC1. IC1 provides the four least-significant bits of the resulting sum, and IC2 provides the four most-significant bits plus a CARRY OUT, which acts as bit-9. This basic circuit can be expanded upwards to accept input words of any desired bit size by taking the CARRY OUT of each successive lower-order stage to the CARRY IN terminal of the next higher-order stage in the chain.

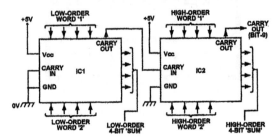

Figure 14.23   Way of using the 74LS83 or 74LS283 as an 8-bit full-adder

The two best known CMOS 4-bit full-adder ICs are the 74HC283, which is a fast CMOS version of the 74LS283, and the 4008B, which has the functional diagram shown in *Figure 14.24*. These two ICs are used in exactly the same way as already shown in *Figures 14.22* and *14.23*.

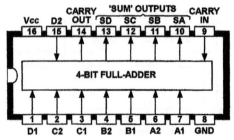

Figure 14.24   Functional diagram of the 4008B 4-bit full-adder IC

Bus Transceiver ICs

In digital electronics, 'bus driver' and 'bus receiver' elements are simply high fan-out buffers or inverters with 3-state outputs, and can be used to either make or break a circuit's input or output contact with a common bus line. *Figure 14.25* illustrates the basic principle, using non-inverting buffers. Thus, in *(a)*, the element acts as a bus driver and allows point 'A' to communicate with 'B' (the bus line) when terminal GAB is high, but is put into the high-impedance 3-state mode when GAB is low. In *(b)*, the element acts as a bus receiver and allows the bus (B) to communicate with point 'A' when GBA is high, but isolates the two points when GBA is low.

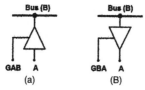

Figure 14.25 3-state buffers used as *(a)* bus driver and *(b)* bus receiver elements

A 'bus transceiver' circuit simply consists of a bus driver and bus receiver element wired in inverse parallel, and allows points A and B to communicate in either direction or to be isolated, as desired. *Figure 14.26* shows a typical bus transceiver circuit, using non-inverting buffers; the L/H element acts as a bus driver and connects A to B when gate terminal NOT-GAB is

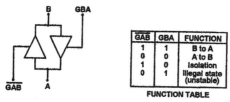

GAB	GBA	FUNCTION
1	1	B to A
0	0	A to B
1	0	Isolation
0	1	Illegal state (unstable)

FUNCTION TABLE

Figure 14.26 Typical non-inverting bus transceiver circuit and Function Table

low, and the R/H element acts as a bus receiver and connects B to A when gate terminal GBA is high. The diagram also shows the circuit's Function Table; thus, the transceiver connects B to A when both gates are high, and connects A to B when both gates are low, and isolates both points when NOT-GAB is high and GBA is low. Note that the GBA-high/NOT-GAB-low condition is an illegal one; it turns both elements on, effectively shorting their outputs and inputs together, and possibly causing latch-up or wild oscillations.

The three best known '74LS'-series bus transceiver ICs are the 74LS242, 74LS243, and 74LS245. The 74LS242 (inverting) and 74LS243 (non-inverting) ICs are 4-line transceivers with similar outlines and pin notations; *Figure 14.27* shows the functional diagram of the 74LS243, which has a Function Table identical to that shown in *Figure 14.26*. The 74LS245 is an 8-line non-inverting transceiver that incorporates logic circuitry that makes it impossible for an 'illegal' input gating state to occur; this IC is housed in a 20-pin DIL package. These three ICs are also available in the '74HC'-series.

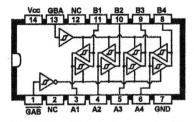

Figure 14.27   Functional diagram of the 74LS243 (or 74HC243) non-inverting 4-line bus transceiver IC

Using Open-drain Outputs

A small proportion of CMOS ICs have open-drain (o.d.) outputs. The cheapest IC of this type is the 40107B Dual 2-input NAND buffer/driver, which (in its popular 8-pin version) has the functional diagram shown in *Figure*

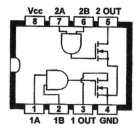

Figure 14.28 Functional diagram of the 40107B Dual 2-input NAND buffer/driver with open-drain outputs (8-pin DIL version)

14.28, and is a useful device for demonstrating 'open-drain' basics. Note that each of this IC's two basic 'elements' effectively consists of a 2-input AND gate with its output connected to the gate of an n-channel MOSFET, which has its drain connected directly to the OUT terminal. This element can be made to function as a 2-input NAND gate by connecting it as shown in *Figure 14.29*, with a load resistor R_L connected between the supply rail and OUT, and with the output voltage taken from the OUT terminal.

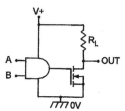

Figure 14.29 Basic way of using a 40107B element as a 2-input NAND gate

Each 40107B element can be made to act as a simple buffer/driver with an o.d. output by shorting its two inputs together, as shown in *Figure 14.30*; this simple

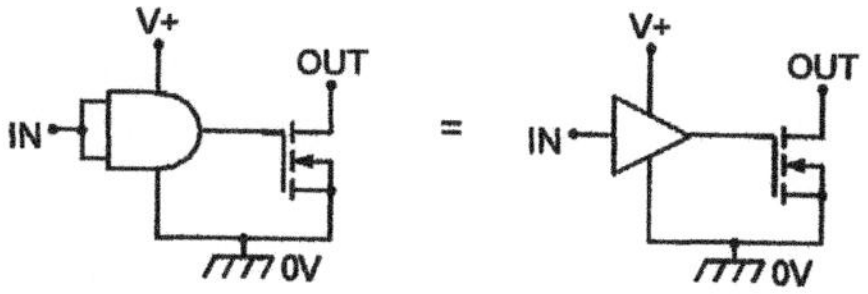

Figure 14.30 Way of using a 40107B element as a simple buffer with o.d output

element can be used in a variety of basic ways, as shown in *Figures 14.31* and *14.32*. Thus, the element can be used as a simple inverter by connecting it as shown in *Figure 14.31*(a), or as an inverting level shifter (which converts input switching levels of V+(1) into output switching levels of V+(2)) by connecting it as shown in *Figure 14.31*(b).

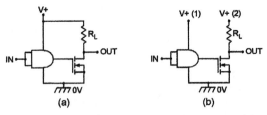

Figure 14.31   Basic ways of using a 40107B element as *(a)* an inverter or *(b)* an inverting level shifter

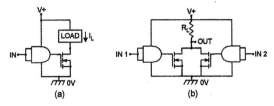

Figure 14.32   Basic ways of using 40107B elements as *(a)* load drivers or *(b)* 'wired NOR' gates

Figure 14.32(a) shows how a 40107B element can be used as a load driver that activates the load when the input is driven to logic-1; the output can sink load currents of up to 100mA when the element is powered from a 15V supply. Finally, *Figure 14.32*(b) shows how two basic elements can be connected in the basic 'wired NOR' gate configuration, in which the two IGFET outputs share a common load (R_L) and the output is thus driven low when either input goes high; in practice, any number of basic elements can be wired in parallel in this fashion to make a wired NOR gate of any desired 'input' size.

Miscellaneous Digital IC Types

All of the most important basic types of current-production digital ICs have now been covered in this volume, but there are a still a few types that have not yet been mentioned, because they are obsolescent or obsolete and may only be found in older equipment, or are unduly expensive or hard to find, etc. Included among these are the following devices:

Parity generators/checkers. Whenever a digital word is transmitted along a severely interference-prone communication link it is possible that the word may be so degraded that its code is changed from that transmitted. Parity generator/checker ICs offer a simple means of checking the truth of a word's code, via an extra 'parity' bit; when the word is to be transmitted, the IC looks at the number of '1's that it contains and (typically) adds an extra '1' bit to it if the number is even, or an extra '0' bit if the number is odd. When the word is received, the IC looks at it and its parity bit and checks them for compatibility (parity); if parity exists, the word is assumed to be pure, and is passed, but if parity does not exist the word is proved to be corrupt, and the receiver system may then request the retransmission of the original word. The parity-checking system is not fool-proof, but does offer a very high level of security.

The best known current-production TTL IC of this type is the 74180 9-bit (8 data bits plus 1 parity bit) parity generator/checker, which uses standard TTL technology. The best known current-production CMOS parity generator/checker ICs are the 40101B and 74HC280 9-bit (8 data bits plus 1 parity bit) and the 4531B 13-bit (12 data bits plus 1 parity bit) types. Each of these ICs can be used to check fewer than its maximum number of input bits by wiring half of the unwanted inputs high and the other half low, or can be used to check more than its maximum number of input bits by cascading appropriate numbers of ICs.

Rate multiplexers. These are a special type of programmable divider, in which the divide-by value equals *B/X*, where *X* is a settable limited-range whole-number value and B is a fixed base number (usually 10, 16 or 64); thus, if B = 10 and *X* is set at 7, the IC will output seven pulses for every ten that are applied at the input and have a divide-by value of 1.4286; if *X* is stepped through numbers 1 to 9, the IC output will provide 1 to 9 output pulses for every ten input pulses, and have divide-by values that step through the values 10, 5, 3.333, 2.5, 2, 1.667, 1.4286, 1.25, and 1.111. ICs of this type are (or used to be) useful in performing arithmetic operations such as multiplication and division, and in A-to-D and D-to-A conversion, etc., but have now been generally superseded by more sophisticated types of LSI device. The best known CMOS IC of this type is the 4527B decade rate multiplier, which has a base number of 10 and can accept any '*X*' number from 1 to 9 (set via a 4-bit address).

Voltage controlled oscillators (VCOs). A VCO is a circuit that generates a good squarewave output waveform that has its operating frequency variable over a wide range via a DC input voltage. The two best known CMOS ICs that provide this action are the 4046B and 74HC4046 'phase-locked loop' (PLL) ICs, and Chapter 9 shows various ways of using the VCO sections of these ICs. Chapter 15 shows some practical ways of using these ICs and their VCOs in the PLL mode.

15   Circuit Miscellany

To conclude this 'Pocket Book', this final chapter presents a miscellaneous collection of useful digital IC circuits that do not readily fit under any of the specific classification headings of earlier chapters. These circuits include alarm-call generators, security alarms, DC lamp controllers, and some unusual designs based on the 4046B 'phase-locked loop' IC, plus a CMOS 6-decade (1Hz to 1MHz) crystal calibrator circuit.

Alarm-call Generators

Inexpensive CMOS logic ICs such as the 4001B can easily be used (in conjunction with one or more transistors) to make a variety of audible-output alarm-call generator circuits. These generators may give monotone, pulsed tone, or warble tone outputs, may give latching, non-latching, or one-shot operation, and can be designed to give output powers ranging from a mere few milliwatts up to about 18 watts. The present section shows only low-power versions of these alarm-call generators; power-boosting circuits are described in a later section.

Figure 15.1 shows a low-power switch-activated 800Hz (monotone) alarm-call generator. Here, two gates of a 4001B are wired as a gated astable multivibrator (see

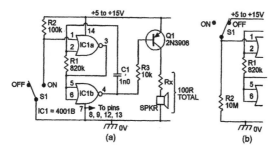

Figure 15.1   Low-power 800Hz monotone alarm-call generator designed for *(a)* n.o. and *(b)* n.c. switch activation

Chapter 9) with its frequency set at 800Hz by R1–C1, and the IC's two other gates are disabled by grounding their inputs. The astable's action is such that it is inoperative, with its pin-4 output high, when S1 is open (pin-1 high), but acts as a squarewave generator when S1 is closed (pin-1 low). Thus, when S1 is open, the astable is inoperative and zero base current is fed to Q1, and the circuit consumes only a small leakage current, but when S1 is closed the astable is operative and feeds a squarewave into the speaker via Q1. The circuit can be activated via normally open (n.o.) switch contacts by using the connections shown in *Figure 15.1(a)*, or by normally closed (n.c.) contacts by using the input connections shown in *Figure 15.1(b)*. In the latter case the circuit draws a standby current of about 1μA via bias resistor R2.

The basic *Figure 15.1* circuit is intended for low-power applications only, and can be used with any speaker in the range 3R0 to 100R and with any supply in the range 5V to 15V. Note that resistor Rx is wired in series with the speaker and must have its value chosen so that the total series resistance is roughly 100R, to keep the dissipation of Q1 within acceptable limits. The actual power output level of the circuit depends on the individual values of speaker impedance and supply voltage that are used, but is no more than a few tens of milliwatts. If desired, the output power can be raised as high as 18 watts by using one of the booster circuits shown later in this chapter.

Figure 15.2 shows a 4001B used to generate a pulsed 800Hz alarm-call signal when S1 is operated. Here, IC1a–IC1b are wired as a low-frequency (6Hz) gated astable multivibrator that is activated by S1, and

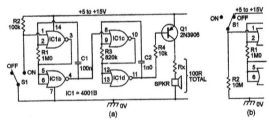

Figure 15.2 Low-power pulsed-tone alarm-call generator designed for *(a)* n.o. and *(b)* n.c switch activation

IC1c–IC1d are wired as an 800Hz astable that is activated by the 6Hz astable. Normally, when S1 is open, both astables are inoperative and the circuit consumes only a small leakage current. When S1 is closed, both astables are activated and the low-frequency astable pulses the 800Hz astable on and off at a 6Hz rate; a pulsed 800Hz squarewave is thus fed to the speaker. The circuit can be activated via an n.o. switch by using the connections shown in *Figure 15.2(a)*, or by an n.c. switch by using the connections of *Figure 15.2(b)*.

Figure 15.3 shows the above circuit modified so that it generates a warble-tone alarm signal that switches alternately between 600Hz and 450Hz at a 6Hz rate. The circuit is basically similar to that just described, except that the 6Hz astable is used to modulate the frequency of the right-hand astable rather than to pulse it on and off. Note that the pin-1 and pin-8 'gate' terminals of both astables are tied together, and the astables are thus both activated directly by the input switch. The circuit can be activated via n.o. switches by using the connections of *Figure 15.3(a)*, or by n.c. switches by using the connections of *Figure 15.3(b)*.

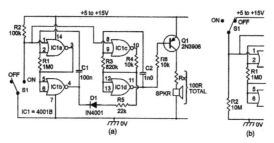

Figure 15.3 Low-power warble-tone alarm-call generator designed for *(a)* n.o. and *(b)* n.c. switch activation

Self-latching Circuits

The alarm-call generator circuits of *Figures 15.1* to *15.3* are all non-latching types which produce an output only while activated by their control switches. By contrast,

Figures 15.4 and *15.5* show circuits that give some form of self-latching alarm-generating action.

The *Figure 15.4* circuit is that of a one-shot or auto-turn-off alarm-call generator. Here, IC1a–IC1b are wired as a one-shot (monostable) multivibrator that can be triggered by a rising voltage on pin-2, and IC1c–IC1d are wired as a gated 800Hz astable multivibrator that is activated by the output of the monostable. Thus, the circuit action is such that both multivibrators are normally inoperative and the circuit consumes only a small leakage current. As soon as S1 is momentarily activated, however, the monostable triggers and turns on the 800Hz astable, which then continues to operate for a preset period, irrespective of the state of S1. At the end of this period the alarm signal automatically turns off again, and the action is complete. The circuit can be retriggered by applying another rising voltage to pin-2 via S1; *Figures 15.4(a)* and *15.4(b)* show alternative switching methods. The alarm duration time is determined by C1, and approximates one second per microfarad of C1 value; periods of several minutes can readily be obtained.

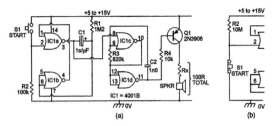

Figure 15.4 One-shot 800Hz monotone alarm designed for *(a)* n.o. and *(b)* n.c. switch activation

Figure 15.5 shows a true self-latching 800Hz switch-activated monotone alarm-call generator. Here, IC1a–IC1b are wired as a manually triggered bistable multivibrator, and IC1c–IC1d as a gated 800Hz astable that is activated via the bistable. The circuit action is such that the bistable output is normally high and the astable is disabled, and the circuit consumes only a small leakage current. When S1 is briefly operated, pin-2 of the IC is pulled high, so the bistable changes state

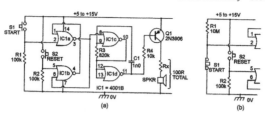

Figure 15.5 Self-latching 800Hz monotone alarm for *(a)* n.o. and *(b)* n.c. switch activation

and its output locks low and activates the 800Hz astable. Once the alarm signal has been so activated, it can only be turned off again by removing the positive signal from pin-2 and briefly closing RESET switch S2, at which point the circuit resets and again consumes only a small leakage current.

Power Boosters

The mean output power of the *Figure 15.1* to *15.5* circuits depends on the individual values of speaker impedance and supply voltage used, but is usually of the order of only a few tens of milliwatts. Using a 9V supply, for example, the output power to a 15R speaker is about 25mW, and to a 100R speaker is about 160mW. If desired, the output powers of these circuits can be greatly increased by modifying their outputs to accept the power booster circuits of *Figure 15.6* or *Figure 15.7*.

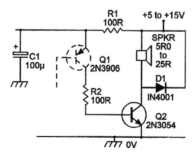

Figure 15.6 Medium-power (0.25W to 11.25W) booster stage

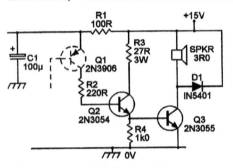

Figure 15.7   High-power (18W) booster stage

In these circuits, R2 is wired in series with the collector of the existing Q1 alarm output transistor and provides base drive to a 1- or 2-transistor booster stage, and the alarm's power supply is decoupled from that of the booster via R1–C1. Note that protection diodes are wired across the speakers of the booster circuits, and prevent the speaker back e.m.f.'s from exceeding the supply-rail voltage.

The *Figure 15.6* booster circuit can be used with any speaker in the range 5R0 to 25R and with any supply from 5V to 15V. The available output power varies from 250mW when a 25R speaker is used with a 5V supply, to 11.25W when a 5R0 speaker is used with a 15V supply. The *Figure 15.7* circuit is designed to operate from a fixed 15V supply and to use a 3R0 speaker, and gives a mean output power of about 18W. Note that, because of transistor leakage currents, the *Figure 15.6* and *15.7* circuits pass quiescent currents of roughly 10µA and 30µA respectively when in the 'standby' mode.

Multitone Alarms

Each of the *Figure 15.1* to *15.5* circuits has a single input switch and generates a unique sound when that switch is operated. By contrast, *Figures 15.8* and *15.9* show a couple of 'multitone' alarm-call generators that each have two or three input switches and which generate a unique sound via each input switch. These circuits are useful in identifier applications, such as in door announcing where, for example, a high tone may be generated via a front door switch, a low tone via a back door switch, and a medium tone via a side door switch.

The *Figure 15.8* circuit is that of a simple 3-input monotone alarm-call generator. Here, two 4001B gates are wired as a modified astable multivibrator, and the action is such that the circuit is normally inoperative and drawing only a slight standby current, but becomes active and acts as a squarewave generator when a resistance is connected between pins 2 and 5 of the IC. This resistance must be less than the 2M2 value of R4, and the frequency of the generated tone is inversely proportional to the resistance value used. With the component values shown, the circuit generates a tone of roughly 1500Hz via S1, 800Hz via S2, and 450Hz via S3. These tones are each separated by about an octave, so each push-button generates a very distinctive tone. The basic circuit generates an output power of only a few tens of milliwatts, but this can be boosted as high as 11.25 W by using the booster stage of *Figure 15.8(b)*.

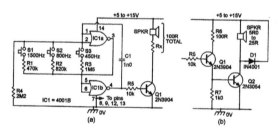

Figure 15.8 3-input multitone alarm *(a)*, plus power-booster stage *(b)*

Figure 15.9 shows a 2-input multitone unit that generates a pulsed-tone signal via S1 or a monotone signal via S2. Here, IC1a–IC1b are wired as a gated 6Hz astable and IC1c–IC1d as a 800Hz astable. The astables are interconnected via D1, and the circuit action is such that the 6Hz astable activates the 800Hz astable when S1 is closed, thus producing a pulsed-tone output signal,

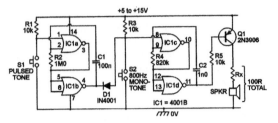

Figure 15.9   2-input multitone alarm

but only the 800Hz astable operates when S2 is closed, thus producing a monotone output signal. This basic circuit generates output powers of only a few tens of milliwatts, but this power can – if desired – be greatly boosted via the circuits of *Figure 15.6* or *Figure 15.7*.

Security Alarms

Most security alarms can be triggered by the opening or closing of one or more sets of electrical contacts. These contacts may be simple push-button switches, hidden pressure-pad switches, or magnetically operated reed-relays, etc., and the alarm systems may give an audible loudspeaker output, an alarm-bell output, or a relay-type output, etc. Such alarms have many uses in the home and in industry. They can be used to grab attention when someone operates a push switch, or when an intruder opens a window or door or treads on a pressure-pad, or when a piece of machinery moves beyond a preset limit and activates a microswitch, etc. Several different types of relay-output security alarm are described in the present section; the simplest of these is shown in *Figure 15.10*.

In *Figure 15.10*, S1 is normally closed and the relay is off, so the circuit consumes zero standby current. When any of the S2 switches are briefly closed, the relay turns on and self-latches via contacts RLA/2 and activates the alarm bell via contacts RLA/1. Once activated, the alarm can be reset by briefly opening S1. Note that any number of n.o. switches can be wired in parallel in the

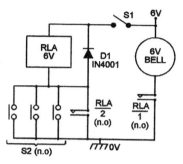

Figure 15.10 Simple security alarm, activated by n.o. switches

S2 position, and the alarm can thus be activated from any desired number of input points.

A basic weakness of the *Figure 15.10* design is that any one of its S2 operating switches can be disabled by simply cutting through the cable that connects it to the main circuit. This snag can be overcome by using the design of *Figure 15.11*, in which all S2 operating switches are n.c. types, wired in series, and the 4001B CMOS gate is wired as an inverter and activates the relay coil via Q1. Normally, with all switches closed, the inverter output is high and Q1 and the relay are off, and the circuit draws a quiescent current of about 1μA via R1. If any of the S2 switches open or have their cables cut, the inverter output rapidly switches low and drives the relay on via Q1, and the relay self-latches via contacts RLA/2 and activates the alarm bell via contacts RLA/1. Note that R2–C1–R3 act as a 'lightning

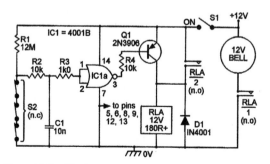

Figure 15.11 Simple security alarm, activated by n.c. switches

suppressor' network that filters out any rapid transient voltages that are induced into the S2 switch cables by the action of lightning. These filter components must be placed as close to the inverter input as possible.

Figure 15.12 shows a simple self-latching burglar alarm circuit that can be activated via any number of parallel-connected n.o. switches (S2) or series-connected n.c. switches (S3); the circuit can also be activated in the non-latching mode at any time via any number of parallel-connected n.o switches (S4). The self-latching alarm action is provided by the two CMOS gates, which are wired in the bistable mode and activate the relay via Q1. This bistable can be SET (turn the relay on) via the S2 or S3 switches and the R3–C1–R4 suppressor network, but is automatically RESET via C2–R6 each time S1 is switched to the 'standby' position.

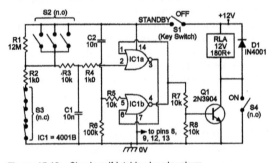

Figure 15.12   Simple self-latching burglar alarm

Figure 15.13 shows the above circuit modified to give auto-turn-off burglar alarm action, causing the relay to turn off after about four minutes of activation. In this case the two CMOS gates are wired in the monostable

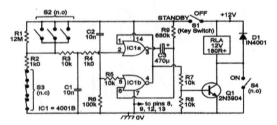

Figure 15.13   Simple auto-turn-off burglar alarm

multivibrator mode, and the four minute delay is determined by the C3–R9 values (C3 must be a low-leakage component).

Next, *Figure 15.14* shows a utility burglar alarm that can be used in many domestic situations. It can be used with any number of series-connected n.c. sensor switches, and gives a self-latching alarm action via relay contacts RLA/1. The circuit action is such that an LED (driven via inverter IC1a and emitter follower Q1) illuminates if any of the n.c. sensor switches are open, but the actual

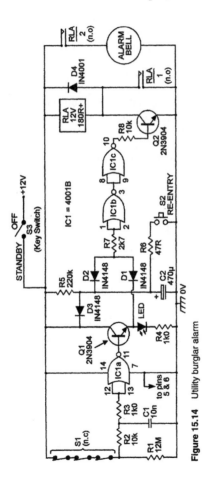

Figure 15.14 Utility burglar alarm

alarm system is automatically disabled (via the C2–R5 time-delay network) for an initial 50 seconds when key-switch S3 is first set to 'standby', thus giving the owner time to reset the switches or leave the house without sounding the alarm; the owner can subsequently re-enter the house without sounding the alarm by reactivating the 50 second delay via hidden external RE-ENTRY switch S2. The actual alarm bell (activated via relay contacts RLA/2) can use the same power supply as the main circuit, or can be separately powered.

Finally, *Figure 15.15* shows how another relay can be wired to the above circuit so that it also gives FIRE and PANIC protection. Here, n.c. push-button switch S4 and n.o. switches S5–S6–RLB/1 are all wired in series with the coil of relay B, and the combination is permanently wired across the supply lines; the n.o. RLB/2 contacts are used to connect the existing alarm generator to the supply lines. Normally, RLB is off, but turns on and self-latches and activates the alarm generator if any of the S5 or S6 switches briefly close. Any number of n.o. PANIC buttons can be wired in parallel with S5, and any number of n.o. thermostats can be wired in parallel with S6. Once RLB has turned on it can be turned off again via S4, which may be in a hidden position.

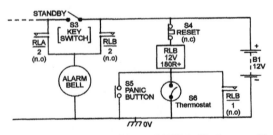

Figure 15.15 Add-on FIRE and PANIC facility for use with *Figure 15.14*

DC Lamp Dimmer

Figure 15.16 shows how a 4001B CMOS IC and a couple of transistors can be used to make a highly

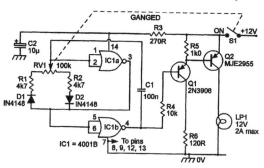

Figure 15.16   12 volt DC lamp dimmer

efficient lamp dimmer that can be used to control the brilliance of the internal lights of any vehicle fitted with a 12V negative-ground electrical system. The two CMOS gates are wired as a 100Hz astable multivibrator that drives the lamp via Q1–Q2 and has its duty cycle or Mark–Space ratio fully variable from 1:20 to 20:1 via RV1, thus enabling the mean power drive to the lamp to be varied from about 5 to 95% of maximum via RV1. Since the period of the 100Hz waveform (10mS) is short relative to the thermal time constant of the lamp, its brilliance can be varied from near zero to maximum with no sign of flicker. Note that ON/OFF switch S1 is ganged to RV1, so the circuit can be switched fully off by simply turning RV1 fully anticlockwise.

Manual Clocking

Chapter 10 showed how simple clocked flip-flop ICs such as the 4013B Dual D-type can be used as counter/dividers. One thing that was not explained was how to clock these counters manually, via a push-button switch. This omission is put right in *Figure 15.17*, which shows how to clock a single 4013B stage via S1. Here, the two 4001B gates are configured as a monostable multivibrator that generates a single 10mS clock pulse each time S1 is operated, thus causing the D-type flip-

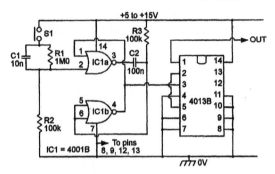

Figure 15.17 Manual clocking of a 4013B counter/divider stage

flop to change state; thus, S1 must be operated twice to make the flip-flop go through a complete ON/OFF or divide-by-2 sequence. Note that the C1–R1 network effectively debounces S1 and helps ensure clean triggering of the monostable circuit.

4046B PLL Circuits

Chapter 9 introduced the 4046B phase-locked loop (PLL) IC and showed how it can be used in voltage-controlled oscillator (VCO) applications. This chip actually contains two phase comparators, a wide-range VCO, a zener diode, and a few other bits and pieces, and is specifically intended for use in PLL applications such as automatic frequency tracking, frequency multiplication, and frequency synthesis, etc. The basic operating principle of the PLL can be understood with the aid of the frequency tracking circuit of *Figure 15.18*. Here, the phase comparator element has two input terminals, one fed from an external input signal and the other from the output of the VCO element. The comparator compares the phase and frequency of the two inputs and generates an output proportional to their difference; this output signal is then smoothed via the low-pass filter network and fed to the control input terminal of the VCO, thus completing the phase-locked feedback loop.

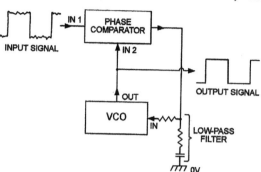

Figure 15.18 Basic PLL frequency tracking circuit

The basic action of the above circuit is such that if the VCO frequency is below that of the external signal the comparator output goes positive and (via the filter network) causes the VCO frequency to increase until both its frequency and phase precisely match (phase lock with) those of the external signal. If the VCO frequency rises above that of the external signal the reverse action takes place, and the comparator output goes low and makes the VCO frequency decrease until it finally locks to that of the external signal. Thus, this circuit causes the VCO signal to automatically phase lock to the external input signal.

Note in the above tracking circuit that the VCO generates a clean (noise-free) and symmetrical output waveform, even if the input signal is noisy and non-symmetrical, and that (because the low-pass filter has a finite time constant) the VCO tracks the MEAN phase and frequency of the input signal. It can thus be used to track and clean up slowly varying input signals, or to track the centre frequency of an FM signal and provide a demodulated signal at the comparator output.

Figure 15.19 shows how the 4046B can be used as a practical wide-range PLL that will capture and track any input signal within the 100Hz to 100kHz (approximate) span range of the VCO, provided that the pin-14 input signal switches fully between the logic-0 and logic-1 levels. Filter R2–R3–C2 is used here as a sample-and-hold network, and its component values determine the settling and tracking times of signal capture. The VCO

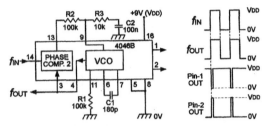

Figure 15.19 Wide-range PLL signal tracker, showing waveforms obtained when the loop is locked

frequency is controlled by C1–R1 and the pin-9 voltage; the VCO span range (and thus the capture and tracking range of the circuit) varies from the frequency obtained with pin-9 at zero volts to that obtained with pin-9 at full supply-rail value.

Figure 15.20 shows a lock detector/indicator that can be used with the above PLL circuit; the circuit's operating principle is moderately complex. Within the 4046B IC, the output of each of the two phase comparators comprises a series of pulses with widths proportional to the difference between the two input signals of the comparator. When the PLL circuit is locked (see *Figure 15.19*) these two outputs are almost perfect mirror images of each other; when the loop is not locked the signals are greatly different. In *Figure 15.20*, these two outputs are fed to the inputs of IC1a, and the circuit action is such that when the loop is locked the IC1a output is permanently low and illuminates LED1 via IC1b, but when the loop is not locked the IC1a output comprises a series of pulses that rapidly charge C1 via D1–R1 and thus drive IC1b output low and turn LED1 off.

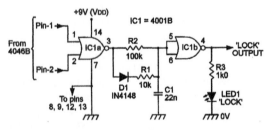

Figure 15.20   PLL lock detector/indicator

Figure 15.21 shows how a PLL circuit can be combined with the above lock indicator to make a precision narrowband tone switch. In this case the maximum VCO frequency is determined by C1–R1, and the minimum by C1 and (R1 + R2); the frequency is variable from about 1.8kHz to 2.2kHz with the component values shown, and the circuit can thus only lock to signals within this frequency range; the circuit output is normally low, but switches high when locked to a suitable input signal.

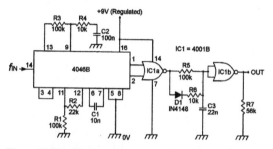

Figure 15.21   Precision narrow-band (1.8kHz to 2.2kHz) tone switch

Frequency Synthesis

One of the most useful applications of the PLL is as a frequency multiplier or synthesizer. *Figure 15.22* shows the basic principle. This circuit is similar to that of the basic

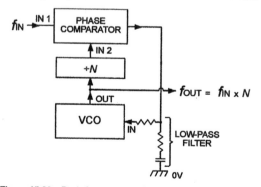

Figure 15.22   Basic frequency synthesizer or multiplier circuit

PLL (*Figure 15.18*) circuit, except for the addition of the divide-by-*N* counter between the VCO output and the phase comparator input. The circuit action is such that the VCO frequency automatically locks to a value at which the divider output frequency matches that of the external input signal, and under this condition the VCO frequency is obviously *N* times the input frequency (where *N* is the counter's division ratio). If the input signal is derived from a precision crystal source, output signals of equal precision can thus be synthesized at any desired multiple frequency by simply using a divider with a suitable *N* value.

Figure 15.23 shows a practical example of a simple frequency synthesizer. It is fed with a precision (crystal derived) 1kHz input signal, and provides an output that is a whole-number multiple (in the range ×1 to ×9) of this signal. The 4017B is used as a programmable divide-by-*N* counter in this simple application, but can easily be replaced by a string of programmable decade 'down' counters of the type described in Chapter 12, to make a wide-range (10Hz to 1MHz) synthesizer.

Finally, *Figure 15.24* shows how the synthesizer principle can be used to make a ×100 frequency pre-scaler that can be used to change a hard-to-measure 1Hz

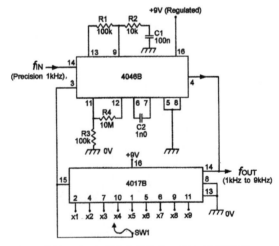

Figure 15.23   Simple 1kHz to 9kHz frequency synthesizer

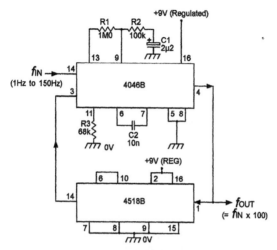

Figure 15.24 A x100 low-frequency pre-scaler

to 150Hz input signal into a 100Hz to 15kHz output signal that can easily be measured on a standard frequency counter. The 4518B IC used in this circuit actually contains a pair of decade counters, and in *Figure 15.24* these are cascaded to make a divide-by-100 counter.

A CMOS Multi-decade Crystal Calibrator

Chapter 11 (*Figure 11.15*) showed how a simple 1MHz TTL crystal oscillator and six TTL decade dividers can be used to make a precision multi-decade 'crystal calibrator' that generates standard frequency/period outputs of 1MHz/1μs, 100kHz/10μs, 10kHz/100μs, 1kHz/1ms, 100Hz/10ms, 10Hz/100ms and 1Hz/1s. To complete this volume, *Figure 15.25* shows an example of a CMOS-based 6-decade calibrator of this type. The circuit is configured to provide symmetrical squarewave outputs on all ranges. These outputs are selected via

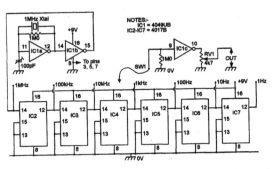

Figure 15.25   CMOS-based 6-decade (1Hz to 1MHz) crystal calibrator

SW1 and buffered via IC1c, and have their amplitudes made fully variable via RV1. The complete circuit consumes less than 2mA when powered from a 9V supply.

Appendix

This appendix contains a full list of the IC types illustrated and described in the text, together with their relevant chapter numbers:

4001B	Quad 2-input NOR gate	Ch. 6
4001UB	Unbuffered Quad 2-input NOR gate	Ch. 6
4002B	Dual 4-input NOR gate	Ch. 6
4006B	18-stage static shift register	Ch. 13
4007UB	Dual complementary pair plus inverter	Ch. 5
4008B	4-bit full-adder	Ch. 14
4011B	Quad 2-input NAND gate	Ch. 6
4012B	Dual 4-input NAND gate	Ch. 6
4013B	Dual D-type flip-flop	Ch. 10
4016B	Quad bilateral switch	Ch. 7
4017B	Decade counter with ten decoded outputs	Ch. 11
4018B	Presettable divide-by-N counter	Ch. 10
4020B	14-stage ripple counter	Ch. 10
4022B	Octal counter with eight decoded outputs	Ch. 11
4023B	Triple 3-input NAND gate	Ch. 6
4024B	7-stage ripple counter	Ch. 10
4025B	Triple 3-input NOR gate	Ch. 6
4026B	Decade counter and 7-segment display driver	Ch. 11
4027B	Dual JK flip-flop	Ch. 10
4028B	BCD-to-decimal decoder	Ch. 13
4029B	Presettable up/down counter	Ch. 12
4033B	Decade counter and display driver	Ch. 11
4040B	12-stage ripple counter	Ch. 10
4042B	Transparent 4-bit data latch	Ch. 13
4046B	Phase-locked-loop IC, with VCO	Ch. 9, 15
4047B	Monostable/astable IC	Ch. 8
4048B	Multifunction expandable 8-input gate	Ch. 6
4049UB	Unbuffered Hex inverter	Ch. 6
4050B	Hex buffer	Ch. 6
4051B	8-channel multiplexer/demultiplexer	Ch. 7
4052B	Dual 4-channel multiplexer/demultiplexer	Ch. 7
4053B	Triple 2-channel multiplexer/demultiplexer	Ch. 7
4060B	14-stage ripple counter	Ch. 10
4063B	4-bit magnitude comparator	Ch. 13
4066B	Quad bilateral switch	Ch. 7
4067B	16-channel multiplexer/demultiplexer	Ch. 7
4068B	8-input NAND gate	Ch. 6
4069UB	Unbuffered Hex inverter	Ch. 6
4070B	Quad EX-OR gate	Ch. 6
4071B	Quad 2-input OR gate	Ch. 6
4072B	Dual 4-input OR gate	Ch. 6
4073B	Triple 3-input AND gate	Ch. 6
4075B	Triple 3-input OR gate	Ch. 6
4077B	Quad EX-NOR gate	Ch. 6
4078B	8-input NOR gate	Ch. 6
4081B	Quad 2-input AND gate	Ch. 6
4082B	Dual 4-input AND gate	Ch. 6
4093B	Quad 2-input Schmitt NAND gate	Ch. 6
4097B	Dual 8-channel multiplexer/demultiplexer	Ch. 7

4098B	Dual monostable	Ch. 8
4099B	8-bit addressable latch	Ch. 14
4502B	Hex 3-state inverter	Ch. 6
4503B	Hex (Dual + Quad) 3-state buffer	Ch. 6
4510B	Decade up/down counter	Ch. 12
4511B	7-segment latch/decoder/LED driver	Ch. 13
4514B	4-bit to 16-way active-high decoder	Ch. 13
4515B	4-bit to 16-way active-low decoder	Ch. 13
4516B	4-bit binary up/down counter	Ch. 12
4518B	Dual synchronous BCD decade counter	Ch. 11
4520B	Dual synchronous 4-bit binary counter	Ch. 11
4522B	Programmable 4-bit decade down counter	Ch. 12
4526B	Programmable 4-bit binary down counter	Ch. 12
4530B	Dual 5-input majority-logic gate	Ch. 6
4543B	Universal 7-segment latch/decoder/driver	Ch. 13
40102B	2-decade BCD down counter	Ch. 12
40103B	8-bit binary down counter	Ch. 12
40106B	Hex Schmitt inverter	Ch. 6
40107B	Dual 2-input NAND buffer/driver (o.d.)	Ch. 14
40192B	Dual-clock decade up/down counter	Ch. 12
40193B	Dual-clock 4-bit binary up/down counter	Ch. 12
74HC00	Quad 2-input NAND gate	Ch. 6
74LS00	Quad 2-input NAND gate	Ch. 6
74HC02	Quad 2-input NOR gate	Ch. 6
74LS02	Quad 2-input NOR gate	Ch. 6
7404	Hex inverter	Ch. 6
74HC04	Hex inverter	Ch. 6
74HCU04	Hex unbuffered inverter	Ch. 6
74LS04	Hex inverter	Ch. 6
74LS05	Hex inverter with o.c. outputs	Ch. 6
7406	Hex inverter with o.c. outputs	Ch. 6
7407	Hex buffer with 30V o.c. outputs	Ch. 6
74HC08	Quad 2-input AND gate	Ch. 6
74LS08	Quad 2-input AND gate	Ch. 6
74LS10	Triple 3-input NAND gate	Ch. 6
74LS11	Triple 3-input AND gate	Ch. 6
74HC14	Hex Schmitt inverter	Ch. 6
74LS14	Hex Schmitt inverter	Ch. 6
74LS20	Dual 4-input NAND gate	Ch. 6
74LS21	Dual 4-input AND gate	Ch. 6
74HC27	Triple 3-input NOR gate	Ch. 6
74LS27	Triple 3-input NOR gate	Ch. 6
74LS30	8-input NAND gate	Ch. 6
74HC32	Quad 2-input OR gate	Ch. 6
74LS32	Quad 2-input OR gate	Ch. 6
7447A	BCD to 7-segment decoder/driver	Ch. 13
74LS47	BCD to 7-segment decoder/driver	Ch. 13
7448	BCD to 7-segment decoder/driver	Ch. 13
74LS48	BCD to 7-segment decoder/driver	Ch. 13
7454	4-wide 2-input AND-OR-INVERT gate	Ch. 6
74LS55	2-wide 4-input AND-OR-INVERT gate	Ch. 6
74LS73	Dual JK flip-flop	Ch. 10
74HC74	Dual D-type flip-flop	Ch. 10
74LS74	Dual D-type flip-flop	Ch. 10
74HC75	Transparent 4-bit data latch	Ch. 13
74LS75	Transparent 4-bit data latch	Ch. 13
74LS76	Dual JK flip-flop with PRESET and CLEAR	Ch. 10

74LS83	4-bit full-adder	Ch. 14
74HC85	4-bit magnitude comparator	Ch. 13
74LS85	4-bit magnitude comparator	Ch. 13
74HC86	Quad EX-OR gate	Ch. 6
74LS86	Quad EX-OR gate	Ch. 6
74LS90	Decade counter with BCD outputs	Ch. 11
74LS93	4-bit JK ripple counter/divider	Ch. 10
74LS107	Dual JK flip-flop	Ch. 10
74121	Standard monostable	Ch. 8
74HC123	Dual retriggerable monostable with CLEAR	Ch. 8
74LS123	Dual retriggerable monostable with CLEAR	Ch. 8
74LS125	Quad 3-state buffer	Ch. 6
74HC132	Quad 2-input Schmitt NAND gate	Ch. 6
74LS132	Quad 2-input Schmitt NAND gate	Ch. 6
74HC138	8-way demultiplexer	Ch. 14
74LS138	8-way demultiplexer	Ch. 14
74HC139	Dual 4-way demultiplexer	Ch. 14
74LS139	Dual 4-way demultiplexer	Ch. 14
74HC151	8-input multiplexer	Ch. 14
74LS151	8-input multiplexer	Ch. 14
74HC153	Dual 4-input multiplexer	Ch. 14
74LS153	Dual 4-input multiplexer	Ch. 14
74HC154	16-way demultiplexer	Ch. 14
74LS154	16-way demultiplexer	Ch. 14
74HC157	Quad 2-input multiplexer	Ch. 14
74LS157	Quad 2-input multiplexer	Ch. 14
74HC190	4-bit decade up/down counter	Ch. 12
74LS190	4-bit decade up/down counter	Ch. 12
74HC191	4-bit binary up/down counter	Ch. 12
74LS191	4-bit binary up/down counter	Ch. 12
74HC192	4-bit decade dual-clock up/down counter	Ch. 12
74LS192	4-bit decade dual-clock up/down counter	Ch. 12
74HC193	4-bit binary dual-clock up/down counter	Ch. 12
74LS193	4-bit binary dual-clock up/down counter	Ch. 12
74LS221	Dual Schmitt-triggered monostable with CLEAR	Ch. 8
74LS240	Octal (Dual Quad) 3-state Schmitt inverter	Ch. 6
74HC243	4-line bus transceiver (non-inverting)	Ch. 14
74LS243	4-line bus transceiver (non-inverting)	Ch. 14
74HC244	Octal (Dual Quad) 3-state Schmitt buffer	Ch. 6
74LS244	Octal (Dual Quad) 3-state Schmitt buffer	Ch. 6
74HC259	8-bit addressable latch	Ch. 14
74LS259	8-bit addressable latch	Ch. 14
74LS260	Dual 5-input NOR gate	Ch. 6
74HC266	Quad EX-NOR gate with o.d. outputs	Ch. 6
74LS279	Quad S–R latch	Ch. 8
74LS283	4-bit full-adder	Ch. 14
74LS365	Hex 3-state buffer	Ch. 6
74LS373	Transparent 8-bit data latch	Ch. 13
74HC375	Transparent 4-bit data latch	Ch. 13
74LS375	Transparent 4-bit data latch	Ch. 13
74HC390	Dual decade (Bi-quinary) counter	Ch. 11
74LS390	Dual decade (Bi-quinary) counter	Ch. 11
74LS393	Dual 4-bit ripple counter/divider	Ch. 10
74HC393	Dual 4-bit ripple counter/divider	Ch. 11
74HC4016	Quad bilateral switch	Ch. 7
74HC4017	Decade counter with ten decoded outputs	Ch. 11
74HC4020	14-stage ripple counter	Ch. 10
74HC4024	7-stage ripple counter	Ch. 10

74HC4040	12-stage ripple counter	Ch. 10
74HC4046	Phase-locked-loop IC, with VCO	Ch. 9
74HC4049	Hex inverter	Ch. 6
74HC4050	Hex buffer	Ch. 6
74HC4051	8-channel multiplexer/demultiplexer	Ch. 7
74HC4052	Dual 4-channel multiplexer/demultiplexer	Ch. 7
74HC4053	Triple 2-channel multiplexer/demultiplexer	Ch. 7
74HC4060	14-stage ripple counter	Ch. 10
74HC4066	Quad bilateral switch	Ch. 7
74HC4075	Triple 3-input OR gate	Ch. 6
74HC4510	Decade up/down counter	Ch. 12
74HC4511	7-segment latch/decoder/LED driver	Ch. 13
74HC4514	4-bit to 16-way active-high decoder	Ch. 13
74HC4515	4-bit to 16-way active-low decoder	Ch. 13
74HC4516	4-bit binary up/down counter	Ch. 12
74HC4543	Universal 7-segment latch/decoder/driver	Ch. 13
74HC40102	2-decade BCD down counter	Ch. 12
74HC40103	8-bit binary down counter	Ch. 12
74HC40192	Dual-clock decade up/down counter	Ch. 12
74HC40193	Dual-clock 4-bit binary up/down counter	Ch. 12
7555	CMOS '555' timer IC	Ch. 9

Index

Individual ICs are listed in the Appendix. They are listed in numerical order (disregarding sub-family letter codes) with the number of the chapter which contains their description and functional diagram.

Printed and bound by CPI Group (UK) Ltd, Croydon, CR0 4YY

03/10/2024

01040433-0002